Improve
Your Maths

Improve Your Maths

A Refresher Course

Jon Curwin
and
Roger Slater

Business Press
Thomson Learning™

Australia • Canada • Denmark • Japan • Mexico • New Zealand • Philippines
Puerto Rico • Singapore • South Africa • Spain • United Kingdom • United States

138980

Improve Your Maths: A Refresher Course

Business Press is a division of Thomson Learning. The Thomson Learning Logo is a registered trademark used herein under licence.

For more information, contact Business Press, Berkshire House, 168–173 High Holborn, London, WC1V 7AA or visit us on the World Wide Web at: http://www.itbp.com

British Library Cataloguing-in-Publication Data
A catalogue record for this book is available from the British Library

ISBN 1-86152-551-6

First edition published 2000 by Thomson Learning

Typeset by Saxon Graphics
Printed in China by L. Rex Printing Co. Ltd
Cover design by Terry Folley
Text design by Gray Publishing, Tunbridge Wells, Kent

Contents

Introduction

This book is designed to help you use numbers more effectively.

Numbers are part of our everyday lives. Numbers can also be an important part of academic studies and the workplace. This book will give you a chance to test your skill, revise areas that you are not sure about and address any gaps in your knowledge. We believe that time spent now on improving your maths will be an investment for the future.

You may have thought that you had seen the last of 'maths' once you left school but now find that you are expected to use all the notation and formulae that you studied for GCSE. Most courses will require some use of mathematics even though the range will vary considerably. Courses as diverse as sociology, psychology, economics and accountancy will all seek to describe the world they are looking at as effectively as they can, using numbers where appropriate.

Many courses will have specific units on statistics or quantitative methods and this book is designed to give you a firm foundation of study. If you are able to spend sufficient time with this book at the beginning of your course, you should be able to approach further study with increased confidence. You can also use this book alongside your studies, checking your maths as you go. If you need a more detailed knowledge, then you can refer to the companion volume *Quantitative Methods for Decision Making* (also by Curwin and Slater), which will take you through most of the quantitative techniques you are likely to use.

This is not the sort of book which you read cover to cover!

Pick it up as and when you need it!

Don't feel you have to finish it in one go!

Often, it helps to re-read something a couple of days or weeks later!

You may already know more than you think!

This book is designed to allow easy use. We suggest that you get to know the structure and the approximate contents of the book so that when necessary, you can find the bits you want quickly. Each Section of the book works to the model shown in Diagram 1. You can find out how much you already know about a topic by doing a quick self-test and then deciding which parts you need to read. You can work your way around the main parts of any section as many times as you wish and, if you want to, you can try the Re-tests in Section Six, just to confirm that you are now competent.

This book is essentially about you and your maths. If you have read this far it probably means that this would be a good time to start improving your maths. Do not be concerned about past results. If you come across a problem you cannot solve, try and identify the cause. If there is a weakness in your basic maths this is the time to resolve it. Improving your maths is like building a house: ensure that the foundations are firm, gradually put the parts into place and understand that all the parts are related in some way.

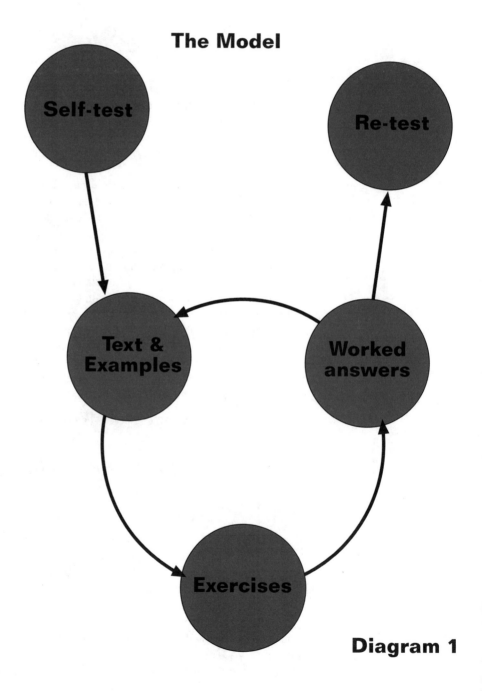

The Model

Diagram 1

Section 1

Numbers

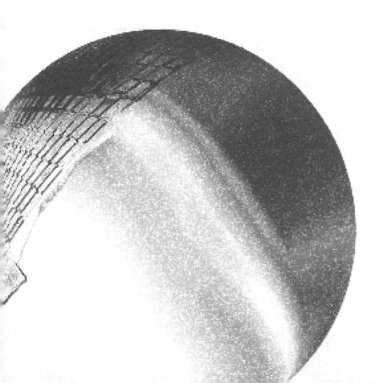

Self-tests on numbers

Here we have a series of questions to help you focus on those parts of Section One where you will need for a good basic understanding of working with numbers. These five self-tests are about different aspects of numbers, and you may find that you are already familiar with and understand many of the ideas involved. When you have tried to answer the questions, check your answers and then make a note of those topics you will need to concentrate on.

Self-test 1: The basics

1 $7 + 5$

2 $15 - 6$

3 $10 - 38$

4 4×7

5 15×6

6 $4 \times (2)$

7 $3 \times (-2)$

8 $(-2) \times (-7)$

9 $6 + 3 \times 2$

10 $(6 + 3) \times 2$

11 $10/2$

12 $-15/3$

Self-test 2: Working with fractions

1 $\frac{1}{2} + \frac{1}{2} + \frac{1}{2}$

2 $\frac{1}{4} + \frac{1}{4}$

3 $\frac{1}{2} + \frac{1}{4} \times 3$

4 $\frac{2}{3} - \frac{1}{6}$

5 $\frac{2}{7} \times \frac{2}{3}$

6 $4\frac{1}{3} \times 1\frac{1}{4}$

7 $\frac{2}{5} \div \frac{1}{2}$

8 $\frac{2}{5} \div 1\frac{1}{2}$

Self-test 3: Decimals and rounding

1 $0.5 + 0.5 + 2$

2 0.8×0.8

3 $0.3 \times 0.6 - 0.1$

4 $0.5 \times \frac{1}{2}$

Round the following number to the given number of places 102.4635571

5 6 decimal places

6 3 decimal places

7 2 decimal places

8 1 decimal place

Self-test 4: Percentages and ratios

1 10% of 50

2 25% off £50 leaves

3 If the price of an item is £35 and it is increased by 10% and then increased by a further 16%, what is the new price?

4 If you get 60% of a cake and your friend gets the rest, the ratio in which the cake is split is?

5 Divide 120 in the ratio 5:3

6 Divide £432 in the ratio 7:2

7 The cost of a joint holiday of £1,650 is to be divided between two families in the ratio 5:6. How much should each pay?

8 The cost of having a ball is to be split between three groups of friends in the ratio 5:4:3. If the cost is £216, how much should each group of friends pay?

Self-test 5: Powers

You should not need a calculator to do this test!

1 $2 \times 2 \times 2 \times 2$

2 3^2

3 $2^2 + 2^3$

4 $2^3 \times 2^4$

5 $(1 + 4^2)$

6 $2^4 \div 2^3$

7 $\sqrt{25}$

8 $\sqrt{(6^2)}$

9 $(3^2) \div (1 + 4 - 2 + 6)$

10 $4^5 \div (4^3 \times 4^2)$

Self-test 1 answers

1 12

2 9

3 −28

4 28

5 90

6 8

7 −6

8 14

9 12

10 18

11 5

12 −5

How many of them did you get right? How many did you end up guessing? If you got more than six right, you've made a good start. Over half of them right! Questions 1 to 5 confirm that you can do the very basics of addition, subtraction and multiplication; if you had any difficulty at all, start at the beginning of Chapter 1. The remaining questions are concerned with the use of brackets and the order of operations (BEDMAS). These topics are dealt with in the second half of Chapter 1. Getting these basics right now will save you lots of problems later on when one simple mistake stops you from getting a complex problem right.

Self-test 2 answers

1 $1\frac{1}{2}$

2 $\frac{1}{2}$

3 $1\frac{1}{4}$

4 $\frac{1}{2}$

5 $\frac{4}{21}$

6 $5\frac{5}{12}$

7 $\frac{4}{5}$

8 $\frac{4}{15}$

Questions 1 and 2 are concerned with adding like terms. Question 3 is a reminder that the order of operations is important (see BEDMAS in Chapter 1). The multiplications in questions 5 and 6 require the multiplication of numerator and denominator. The divisions in questions 7 and 8 require the dividing fraction to be turned upside down before you multiply. These topics are revised in Chapter 2.

Self-test 3 answers

1 3

2 0.64

3 0.08

4 0.25 or $\frac{1}{4}$

5 102.463557

6 102.464

7 102.46

8 102.5

Questions 1 to 4 are concerned with using the basic mathematical operations with decimal numbers. You should now be sure that you can correctly answer this type of question. Questions 5 to 8 are a reminder that calculations can easily produce answers with many decimal places. The importance of the additional decimal places quickly diminishes and it does become necessary to choose how your result should be reported. You will need to remember to round up or down according to the remaining decimal digits.

Self-test 4 answers

1 5

2 £37.50

3 £44.66

4 3:2

5 75 and 45

6 336 and 96

7 750 and 900   **8** 90, 72 and 54

Questions 1 to 3 are concerned with the use of percentages. Many people prefer to work with the language of percentages rather than fractions, and it is important to understand percentage increases and decreases and to be able to make the calculations correctly. The remaining questions are concerned with the use of ratios and how to use ratios to correctly share out a particular total amount.

Self-test 5 answers

1 16   **6** 2

2 9   **7** 5

3 12   **8** 6

4 128   **9** 1

5 17   **10** 1

How did you get on with this test? If you got 9 or 10 right, you probably know most of the things in this section. If you got more than 6, a quick read through may be all you need. For the rest of us, reading through Chapter 5 will prove a useful reminder of the ideas associated with the use of powers.

You are now in a position to decide which parts of Chapters 1–5 you need and want to work through to improve your understanding of numbers.

CHAPTER 1

The basics

We'll start off with a few things which you will need.

This section is not in the least concerned with doing complicated mental arithmetic using large numbers. If you are using big numbers then you will use a calculator or a spreadsheet. Here we are only going to look at a few calculations using small numbers to make sure that we can all agree on certain ways of doing things. Not only will this make sense for calculations in your head, but it will also make it easier to use spreadsheets. It is often a good idea to have some idea of what size an answer should be, so that you can tell if a calculator or spreadsheet has given an incorrect answer.

Addition

Adding numbers together is a concept we meet early in our education, and with a little care, most of us can do this most of the time. Take the sum:

$$3 + 4 = 7.$$

Here we are adding a further 4 to a number 3. Look at Figure 1.1 and you can see how the two parts give an answer of 7 for this sum.

Figure 1.1

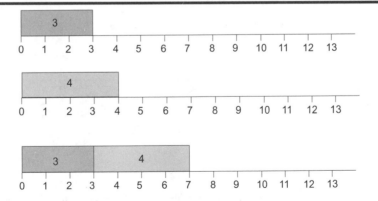

See how we are using the ruler to find the result of the addition.

Subtraction

With subtraction you can still think about two line segments, but one is going the opposite way to the other, so for example

$$6 - 4 = 2$$

We can look at Figure 1.2, again using the ruler as a way of finding the result of the calculation.

Figure 1.2

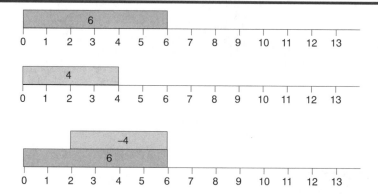

You could even think of this as adding two numbers together, it just happens that one of them is negative.

These ideas will work quite well whilst we are dealing with small numbers and when the result we get is positive. If the number that we are subtracting is bigger than the other number, then, to understand the process, we need to introduce numbers to the left of the zero on our ruler. Looking at Figure 1.3 this will then allow us to read off the result of the subtraction, for example with $4 - 7 = -3$:

Figure 1.3

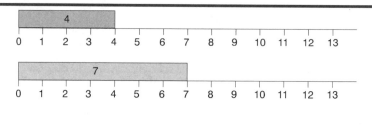

Multiplication

When we multiply we are using the same number several times, so we can think of multiplication as a quick form of addition. For example, take three times two, we can still work with the blocks we used earlier. From Figure 1.4, we can see that:

$$2 + 2 + 2 = 3 \times 2 = 6$$

When one of the numbers which we are multiplying is negative (has a minus sign), then the result will be negative. If both of the numbers are negative, then the result will be positive. In fact, when we are multiplying a series of numbers together, some

Figure 1.4

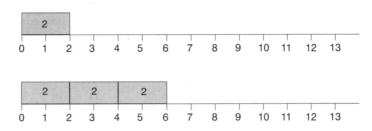

of which are positive, and some of which are negative, then, if there are an even number of negative numbers, the result will be positive, and if there are an odd number of negative numbers, the result will be negative.

You also need to remember that:

any number multiplied by one stays the same;
any number multiplied by zero is equal to zero.

Division

Division is basically the opposite of multiplication. In multiplication we have so many of a number, e.g. 3 2s to get 6. In division, we find out how many of one number we need to make another number. So if we take the number 12, then we can find that it takes 6 2s to make 12. For the number 8, see Figure 1.5.

Figure 1.5

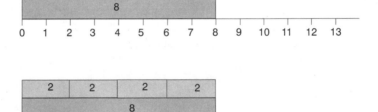

Sometimes division does not give you an exact answer. If you try to find how many 3s you can get into 11, you find that you get 3, with 2 left over. Again, you can see this in Figure 1.6.

Figure 1.6

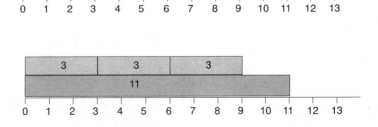

You also need to remember that:

when either of the numbers is negative, the result is negative;
when both numbers are negative, the result is positive.

Brackets   When the expression contains brackets, you need to work out the calculation in the brackets first, and then do the rest. (Working out the result of an expression is often called 'evaluating the expression'.) For example,

$$(3 + 2) \times 6 = 5 \times 6 = 30$$

A Rule – BEDMAS   Looking at how to deal with brackets suggests the need to consider some rules about which things you do first when you evaluate an expression. You may have come across this before, although it is presented in several different ways. Technically the rule is called the order of operations, and can be summarized as:

B	Brackets
E	Exponentiation
D	Division
M	Multiplication
A	Addition
S	Subtraction

We have dealt with most of these operations, except for 'exponentiation', which comes in Chapter 5.

Following this rule, we can now evaluate fairly complicated expressions, and we also have a useful guide when it comes to putting formulae into spreadsheets (see Section Five). For example, if we take this expression:

$$(4 + 3) \times 5 - 3 \times (-2 + 7 + 3)/4$$

the first thing to do is to work out the results of the brackets:

$$7 \times 5 - 3 \times 8/4$$

then we divide:

$$7 \times 5 - 3 \times 2$$

then we do the multiplications:

$$35 - 6$$

and since there are no additions, we then do the subtraction, to get the final result:

29

With a little bit of practice you will be able to do several of these steps in one go without having to write down the intermediate steps, but if you are ever in doubt about a result, then go back to writing down each stage of the calculation.

Exercises

1 12 + 6 −4 × 2

2 49 + 12 × 3 − 31

3 2 × 5 − 7 + (4 + 7) × 3

4 −9 +5 × (−2) + 6

5 (12 + 6)/3

6 (12 − 15) × (−4) + 6

7 (4 − 14)/ (5 − 3)

8 6 × 3 + 4 − 5 + 8/4

9 (5/2) × 3

10 − 3 − 4 × (−2) + 1 − 7/4

Worked answers

1 12 + 6 − 8   *add first*
 18 − 8
 10

2 49 + 36 − 31   *add first*
 85 − 31
 54

3 2 × 5 − 7 + 11 × 3   *multiply first*
 10 − 7 + 33   *now addition*
 43 − 7
 36

4 −9 −10 + 6
 −9 −4
 −13

5 18/3
 6

6 (12 − 15) × (−4) + 6   *brackets first*
 (−3) × (−4) + 6   *minus times minus is positive*
 12 + 6
 18

7 (4 − 14)/ (5 − 3)   *brackets first*
 −10/2
 −5

8 6 × 3 + 4 − 5 + 8/4   *division first*
 18 + 4 − 5 + 2   *gather together all of the positive terms*
 24 − 5   *and then all of the negative terms*
 19

9 (5/2) × 3
 2.5 × 3
 7.5

10 − 3 − 4 × (−2) + 1 − 7/4   *division and multiplication here*
 −3 + 8 +1 − 1.75   *collecting positive and negative terms*
 9 − 4.75   *finally subtraction*
 4.25

You are now ready to go on to Chapter 2, or you might want to check your understanding of the basics by working through the Re-tests in Section Six.

CHAPTER 2

Working with fractions

Unfortunately you cannot describe the world just using whole numbers. You must be able to deal with fractions; be able to easily swop between decimals and fractions; and quote numbers as percentages. You might even want to divide things in a certain ratio, or give odds on an event happening. In this chapter we will look at the first of these ideas and explain how you might use them. Inevitably we have to assume that you have a basic understanding of the material in Chapter 1!

What is a fraction?

A fraction is a way of expressing part of the whole thing – often just part of one. Usually the top line of a fraction is called the **numerator,** and the bottom line is called the **denominator**. If you divide some money equally between two people, then each one gets half ($\frac{1}{2}$) of the whole. If there were 4 people, then the fraction would be a quarter($\frac{1}{4}$). In terms of **addition** and **subtraction** fractions can be treated in much the same way as whole numbers. Sometimes it will be relatively obvious what the answer is, for example if you add two halves together, you might guess the answer is one, or four halves would give you an answer of two.

Where it is not so obvious, we need a method for adding or subtracting these fractions. To do this we try to find a number which all of the bottom lines of the fractions will divide into (called a **common denominator**). For example if we wanted to add $\frac{1}{2}$ and $\frac{1}{3}$ we need to look for a number that 2 and 3 will both divide into. The smallest such number is 6. Now, $\frac{1}{2}$ is equal to $\frac{3}{6}$ and $\frac{1}{3}$ is equal to $\frac{2}{6}$, so adding we get:

$$\frac{1}{2} + \frac{1}{3} = \frac{3}{6} + \frac{2}{6} = \frac{(3+2)}{6} = \frac{5}{6}$$

As a further example:

$$\frac{5}{6} - \frac{3}{5} = \frac{25}{30} - \frac{18}{30} = \frac{7}{30}$$

Multiplying fractions is just a matter of multiplying both (or all) of the top lines and putting the result over the multiplication of both (or all) of the bottom lines. For example:

$$\frac{1}{4} \times \frac{3}{7} = \frac{1\times3}{4\times7} = \frac{3}{28}$$

And another example:

$$\frac{3}{5} \times \frac{5}{9} = \frac{3\times5}{5\times9} = \frac{15}{45}$$

This can be simplified by dividing the top and bottom lines by 15, to give $\frac{1}{3}$

Division of fractions uses the fact that you can turn the dividing fraction upside-down and then multiply. For example if you wish to divide $\frac{1}{2}$ by $\frac{1}{4}$ then you get

$$\frac{1}{2} \div \frac{1}{4} = \frac{1}{2} \times \frac{4}{[1]} = \frac{4}{[2]} = 2$$

This can be easily understood as a quarter goes into a half twice!

As another example:

$$\frac{3}{4} \text{ divided by } \frac{5}{7} = \frac{3}{4} \times \frac{7}{5} = \frac{21}{20}$$

Exercises

1 $\frac{1}{4} + \frac{1}{4} + \frac{1}{4}$

2 $\frac{1}{5} + \frac{1}{6} + \frac{1}{12}$

3 $\frac{1}{4} - \frac{1}{8}$

4 $1\frac{1}{2} - 2\frac{1}{4}$

5 $\frac{1}{3} \times \frac{1}{4}$

6 $1\frac{1}{3} \times \frac{3}{5}$

7 $\frac{2}{3}$ divided by $\frac{1}{6}$

8 $\frac{5}{8}$ divided by $1\frac{1}{4}$

Worked answers

1 $\frac{1}{4} + \frac{1}{4} + \frac{1}{4}$
$\frac{(1 + 1 + 1)}{4} = \frac{3}{4}$ *common denominator is obviously 4*

2 $\frac{1}{5} + \frac{1}{6} + \frac{1}{12}$
$\frac{12}{60} + \frac{10}{60} + \frac{5}{60} = \frac{27}{60}$ *common denominator is 60*

3 $\frac{1}{4} - \frac{1}{8} = \frac{(2-1)}{8} = \frac{1}{8}$

4 $\frac{3}{2} - \frac{9}{4} = \frac{(6-9)}{4} = -\frac{3}{4}$ *common denominator is 4*

5 $\frac{1}{3} \times \frac{1}{4} = \frac{1}{12}$ *just multiplying top and bottom lines*

6 $\frac{4}{3} \times \frac{3}{5} = \frac{12}{15} = \frac{4}{5}$

7 $\frac{2}{3} \times \frac{6}{1} = \frac{12}{3} = 4$ *turning the $\frac{1}{6}$ upside down*

8 $\frac{5}{8} \times \frac{4}{5} = \frac{20}{40} = \frac{1}{2}$

You are now ready to go on to Chapter 3, or you might want to check your understanding of working with fractions by working through the Re-tests in Section Six.

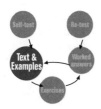

CHAPTER 3

Decimals and rounding

Whilst fractions are often an easy way to think of parts of a whole, most calculations which we see use **decimals**. When you study probability you will use fractions to begin thinking about the topic, but fairly quickly will have to move on to using decimals as the results begin to look more complex. Most of the sums you will need to do, you will do by using a calculator or a computer, and as they work in decimals rather than fractions, we need to know a little bit about decimals.

Decimals

You get a decimal answer by taking your fraction and actually doing the calculation. So, if we take $\frac{1}{2}$, then this is 1 divided by 2. Well, 2 into 1 won't go, but if we put a 0 behind the 1, we have 2 into 1.0 (treat it like 10), and we get 0.5 (since the answer must be less than 1). When we start to do sums with decimals, we have to remember always to note the number of decimal places we are using. For example, if we have 0.5 plus 0.25, we should line up the decimal points, and then add as usual:

$$\begin{array}{r} 0.50 \\ + \ \underline{0.25} \\ 0.75 \end{array}$$

Note that we have put in a 0 to make 0.5 into 0.50, to make each decimal have the same number of digits – **this does not change the value in any way**.

Look at the two following examples:

$$\begin{array}{r} 3.875 \\ + \ \underline{0.445} \\ 4.320 \end{array}$$

$$\begin{array}{r} 4.98421 \\ - \ \underline{2.54293} \\ 2.44128 \end{array}$$

When we multiply and divide, then the need to check the number of decimal places becomes even more important. When we multiply two decimals which are **both less than one**, then the result is **smaller** than either of them. Think about 0.5 times 0.5, and put it back into words – we want the result of finding a half of a half. The answer will be a quarter, or 0.25. You can see that the result has two decimal places, whilst the original numbers each had one. A general rule is that when we multiply two decimals together, the number of decimal places in the result will be the sum of the number of decimal places in the two original numbers.

Another example:

$$0.25$$
$$\times\ \underline{0.5}$$
$$0.125$$

Here we have done the sum – 5 times 25 and got the result 125, counted up the number of decimal places we need, $2 + 1 = 3$, and thus got the result. Looking at a further example:

$$1.5$$
$$\times\ \underline{2.5}$$
$$3.75$$

Finally, we can look at division. This is treated in much the same way as multiplication; but if in doubt, think back to some simple examples, and even put them into words. For example, 2.5 divided by 0.5 is equivalent to 'how many halves are there in two and a half?' We know the answer is five!

Another example would be:

$$5.2 \div 1.3 = 4$$

Rounding

Some calculations lead to exact answers, such as 5 divided by 2 gives 2.5. However, in other cases, the result may have either a very large, or an infinite number of decimal places 3.141592654 (pi (or π) for those who remember school geometry). In such circumstances we often don't want to quote all of the decimal places, since this just confuses the people we are showing the answer to. Therefore we round the result to a particular number of decimal places. As an example, if you quote an amount of money in pounds, you only give two decimal places, since this represents pence.

As a general rule, you should do the calculations using as many decimal places as possible, and only round when you quote the result. As a guide you only use as many decimal places as make sense in the context (such as the example of money quoted above). The number of decimal places to quote will depend on who the result is for. For example, if someone wants to know how hot tomorrow is likely to be, then saying 14°C is quite sufficient. If, however, you are monitoring a scientific experiment, then a more exact measurement of temperature will be required.

To round decimal places we first decide on the number of places required. We then consider the next decimal place along to the right.

If this figure is **4 or less**, then we can ignore the remaining decimal places, for example:

4.9824863 to 3 decimal places is 4.982

as the fourth decimal digit is less than 5.

If the next digit is **5 or more**, then we increase the final digit to be included by 1 and then ignore the remaining decimal places, for example:

4.9824863 to 4 decimal places is 4.9825

as the fifth decimal digit is 5 or more.

Exercises

1 0.2 + 0.5

2 1.24 + 0.742

3 0.34 + 1.257 + 0.469

4 0.5 − 0.3

5 0.4 − 0.7

6 1.456 − 0.642

7 3 × 0.446

8 0.3 × 0.6

9 3 × 0.6

10 5 / 0.2

11 6.3 / 0.1

12 3.2 / 0.25

Round 75.9326489 to

13 6 decimal places

14 4 decimal places

15 2 decimal places

16 1 decimal place

Worked answers

1
```
    0.2
+   0.5
    0.7
```

2
```
    1.240
+   0.742
    1.982
```
making this 3 digits

3
```
    0.340
+   1.257
+   0.469
    2.066
```
making this 3 digits

4
```
    0.5
−   0.3
    0.2
```

5
```
    0.4
−   0.7
−   0.3
```
negative since 0.7 is bigger than 0.4

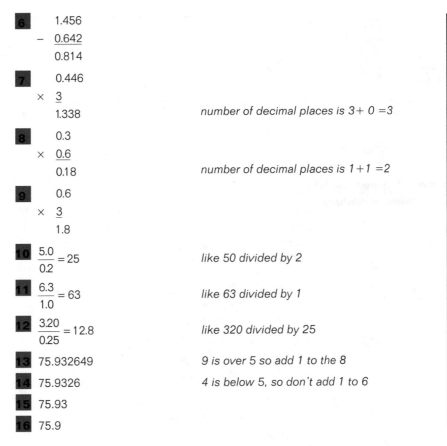

6   1.456
    −   0.642
        0.814

7   0.446
    ×   3
        1.338   *number of decimal places is 3+ 0 =3*

8   0.3
    ×   0.6
        0.18   *number of decimal places is 1+1 =2*

9   0.6
    ×   3
        1.8

10 $\dfrac{5.0}{0.2} = 25$ *like 50 divided by 2*

11 $\dfrac{6.3}{1.0} = 63$ *like 63 divided by 1*

12 $\dfrac{3.20}{0.25} = 12.8$ *like 320 divided by 25*

13 75.932649   *9 is over 5 so add 1 to the 8*

14 75.9326   *4 is below 5, so don't add 1 to 6*

15 75.93

16 75.9

You are now ready to go on to Chapter 4, or you may want the check your understanding of decimals and rounding by working through the Re-tests in Section Six.

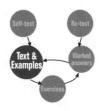

CHAPTER 4
Percentages and ratios

When you have to tell other people about your results it is often easier to quote **percentages**. Most people seem to find percentages easier to understand than fractions or decimals (although some anecdotal evidence suggests that their level of understanding is actually no higher). A percentage can be converted into a fraction just by putting the number over 100, so

$$10\% = \frac{10}{100} \text{ or } \frac{1}{10}$$

For example, if you need to find 20% of 500, then:

$$\frac{20}{100} \times 500 = 20 \times 5 = 100$$

Another example: take a 10% discount off £89. To do this we first find 10% of £89 and then do the subtraction:

$$\frac{10}{100} \times £89 = £8.90$$

$$£89 - £8.90 = £80.10$$

To find a percentage, we can take the number and divide it by the total amount; this gives us a fraction, which we can then multiply by 100 to get a percentage. For example, if we want to find 30 as a percentage of 200, then:

$$\frac{30}{200} \times 100 = 15\%$$

As another example, consider 75 as a percentage of 250:

$$\frac{75}{250} \times 100 = 30\%$$

Ratios

When inputs, payouts, etc. are divided up, although they could be expressed as fractions, decimals or percentages, some people prefer to talk about the **ratio** in which the total amount is divided. To decide the actual shares when a total is divided according to a ratio:

add the ratio numbers together,
then divide the total amount to be split up by that number,
then work out the shares by multiplication.

For example, if we divide the total £200 in the ratio 3:2, we have the following calculations:

$$3 + 2 = 5$$

$$£200/5 = £40$$

So, the actual amounts are:

$$3 \times £40 = £120, \text{ and } 2 \times £40 = £80$$

Another example would be to divide £1000 between three people in the ratio 3:2:5

$$3 + 2 + 5 = 10$$

$$£1000/10 = £100$$

So the amounts are : £300: £200: £500

Exercises

1 10% of 20

2 25% of 50

3 15% off £100

4 If the price of an item is £18, and it then rises by 20%, what is the new price?

5 Express 18 as a percentage of 90

6 Divide 60 in the ratio 6:4

7 Divide 900 in the ratio 5:3:1

8 £1,000 is divided between Anne and Jon in the ratio 7:3. How much does Jon get?

9 Two companies agree to divide the costs of a joint venture in the ratio 3:2. Total cost is £10,000. How much does each pay?

Worked answers

1 10% = 1/10, so 1/10 × 20 = 2

2 25% = 25/100
    = 1/4   *cancelling down*
    so 1/4 × 50 = 12.5

3 15% = 15/100 so 15/100 × 100 =£15 off, leaving £85

4 20% = 20/100 =1/5
1/5 × 18 = £3.60,
so £18 + £3.60 = £21.60   *the new price*
Alternatively, if you say that
the old price was 100%, then
the new price is
100% + 20% = 120% = 1.2,   *higher than the old price*
so 1.2 × £18 = £21.60

5 18/90 × 100 = 20%   *putting the number over the total*

6 6 + 4 = 10   *adding up the ratio numbers*
60/10 = 6,
so the amounts are 6 × 6 = 36, and 4 × 6 = 24

7 5 + 3 +1 = 9   *adding up the ratio numbers*
900/9 = 100,
so 5 × 100 = 500; 3 × 100 = 300; and 1 × 100 = 100

8 7 + 3 = 10
£1,000/10 = £100
so Jon gets £300 (and Anne gets £700)

9 3 + 2 = 5
£10,000/5 = £2,000
so, the first company pays £6,000 and the second pays £4,000

You are now ready to go on to Chapter 5, or you may wish to check your understanding of percentages and ratios by working through the Re-tests in Section Six.

CHAPTER 5

Powers

In some circumstances we end up multiplying the same number by itself many times (this often happens if you study probability) and we need a short-hand form of recording this.

Powers

When you multiply the same number by itself many times, you can use **powers** (a bit like multiplication being a short-hand form of addition). Once we use powers to represent this multiplication, then we need a notation to deal with them. The power is the number of times the number is multiplied, so:

$$2^3 = 2 \times 2 \times 2 = 8$$

Another word used for this process is **exponentiation**.

When we have a number raised to a power, and that result is then multiplied by the **same number** raised to some other power, then we can **add** the powers together. This process is sometimes called **combining powers**. For example:

$$2^3 \times 2^2 = (2 \times 2 \times 2) \times (2 \times 2) = 2^{3+2} = 2^5 = 32$$

Another example:

$$5^4 \times 5^3 = 5^{4+3} = 5^7 = 78125$$

In the same sort of way, if the numbers are divided, then we subtract the powers. For example:

$$3^4 \div 3^2 = 3^{4-2} = 3^2 = 9$$

Another example:

$$10^5 \div 10^3 = 10^{5-3} = 10^2 = 100$$

Where we have a number raised to a power, which is then raised to a power, we can multiply the powers. For example:

$$(3^2)^4 = 3^{2 \times 4} = 3^8 = 6561$$

And again:

$$(7^3)^2 = 7^{3 \times 2} = 7^6 = 117649$$

These results are useful when we are dealing with the same number raised to a power in both cases. Where different numbers are involved, obviously the above doesn't apply. For example:

$$5^3 \times 2^4 = 125 \times 16 = 2000$$

There are **special results** which you will need to be aware of in this context. (*They will also help if you go on to look at various aspects of maths or statistics later.*)

Any number raised to the power zero must be equal to 1

Look at this example:

$16 \div 16 = 1$; fairly obviously.

But $16 = 4^2$, so:

$16 \div 16 = 4^2 \div 4^2 = 4^{2-2} = 4^0 = 1$

Any number raised to the power one must remain the same

Look at this example:

$27 \div 9 = 3$; as you should know.

But $27 = 3^3$, and $9 = 3^2$, so

$27 \div 9 = 3^3 \div 3^2 = 3^{3-2} = 3^1 = 3$

Negative powers mean we are dividing by the number raised to that power

Look at this example:

$4 \div 32 = 1/8$

but $4 = 2^2$, and $32 = 2^5$, so

$4 \div 32 = 2^2 \div 2^5 = 2^{2-5} = 2^{-3}$

We know that $2^3 = 8$, so $1/8 = 1/2^3$

Therefore $2^{-3} = 1/8$

One raised to any power is one

(Something you may have guessed on the way through.)

$$\left(\frac{15}{15}\right)^4 = 1^4 = 1 \times 1 \times 1 \times 1 = 1$$

Fractional powers

The square root of a number is a number that, when it is multiplied by itself, gives the original answer. (The reverse of squaring a number.) For example, since $2 \times 2 = 4$, then the square root of 4 must be 2. (However, since $-2 \times -2 = 4$, then -2 is also a square root of 4). By the same logic, if $9 \times 9 = 81$, then the square root of 81 must be 9 (and also -9). What is the same about these two sums?

$$4 = 2^2, \text{ and the square root is } 2^1$$
$$81 = 3^4, \text{ and the square root is } 3^2$$

To get a square root we halve the power. In other words, the power representing a square root is $\frac{1}{2}$. We could write:

$$\sqrt{16} = 16^{\frac{1}{2}} = (2^4)^{\frac{1}{2}} = 2^{4 \times \frac{1}{2}} = 2^2 = 4$$

A cube root is a number that can be raised to the power 3 to get the required answer. For example:

$$\sqrt[3]{64} = 4, \text{ since } 4^3 = 4 \times 4 \times 4 = 64$$

from this, a cube root could be written as a power of 1/3

$$64^{\frac{1}{3}} = 4^{3 \times \frac{1}{3}} = 4^1 = 4$$

Exercises

1 2^3

2 $2^3 \times 2^4$

3 $3^2 \times 3^2$

4 $2^2 + 2^3$

5 $\sqrt{256}$

6 $(3^2)^2$

7 $\sqrt{(64)}$

8 $\sqrt[3]{(729)}$

Worked answers

1 $2^3 = 2 \times 2 \times 2 = 8$

2 $2^3 \times 2^4 = 2^{3+4} = 2^7 = 128$ *simply adding the powers*

3 $3^2 \times 3^2 = 3^{2+2} = 3^4 = 81$

4 $2^2 + 2^3 = 4 + 8 = 12$ *a trick question*

5 $\sqrt{256} = \sqrt{2^8} = 2^{8 \times \frac{1}{2}} = 2^4 = 16$ *multiplying powers*

6 $(3^2)^2 = 3^{2 \times 2} = 3^4 = 81$

7 $\sqrt{(64)} = 2^{6 \times \frac{1}{2}} = 2^3 = 8$

8 $\sqrt[3]{(729)} = 3^{6 \times \frac{1}{3}} = 3^2 = 9$ *it helps if you realize 729 = 3^6*

You are now ready to go on to next section, but, if you have not already done so, it would be good idea to go to Section Six to check your understanding of numbers.

Section 2

Calculators

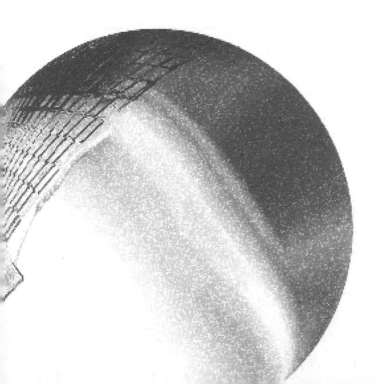

Self-test on using calculators

This test is simply designed to ensure that you can use the basic functions on your calculator. Since we do not know the exact type of calculator which you have, you may want to refer to your manual if your answers do not match ours. You may also find that your calculator can perform much more complex sums than those below.

Self-test

Evaluate each of the following using your calculator:

1 $100 + 527 + 93 + 14 =$

2 $47 - 32 + 5 - 98.5 - 3.1 + 4.03 =$

3 $27 - 4 + 348.3/7 + 4 =$

4 $(27 - 4 + 348.3)/(7 + 4) =$

5 $45 \times 6 \times 0.34 =$

6 $11\% \text{ of } 327 =$

7 $\dfrac{8 \times 453 - 62 \times 9}{8 \times 724 - (62)^2} =$

8 $\dfrac{89}{7} - 0.3483 \times \dfrac{58}{7} =$

9 $\sqrt{(7 \times 4 + 3 \times 14)} =$

10 $\dfrac{147}{(42 \times 729)} =$

Self-test answers

1	734	**6**	35.97
2	−77.57	**7**	1.573 921 971 253
3	76.757 142 857 14	**8**	9.828 371 428 571
4	33.754 545 454 55	**9**	8.366 600 265 341
5	91.8	**10**	0.004 801

How many of them did you get right? The first three questions just look at basic arithmetic – the sort of thing we covered in Section One. As you progressed through the test, the questions became more complex, and if you had difficulties, you may wish to read Chapter 6 to help you get the most out of your calculator. (Remember that your calculator came with an instruction book, and it is worth looking at it to see how it can save you time when doing this sort of calculation.)

CHAPTER 6

Using calculators

Calculators vary tremendously in what they can do, but all calculators will carry out the basic functions of addition, subtraction, multiplication and division. Yours will probably also be able to find square roots. You should find that if you completely master the use of your calculator, it will save you a large amount of work and time. However, you do need to look at what you are doing, since it is very easy just to put numbers blindly into the calculator and accept the result (even if it does not make sense). In this part we are concerned with just the basic functions that are available on most calculators. You may need to remind yourself of the BEDMAS rule given in Chapter One.

One aim you should have in using your calculator is to avoid writing down intermediate answers. To do this you need to plan the way in which you enter the problem into the machine, and then all you need to write down is the final answer. The reason for doing this, apart from it saving you work (!), is to ensure that the intermediate results use as many decimal places as possible, as this will affect the accuracy of the final result.

Basic functions

The four basic arithmetic functions are controlled by the usual four buttons:

$+$		Addition
$-$		Subtraction
$\times$	or $*$	Multiplication
$\div$	or $/$	Division

and we would anticipate you having little difficulty in actually pressing the appropriate one when required. For example, if you need to work out $23 \times 4 \times 0.875$ you would switch the calculator on, then

Type 23	Press	$\times$
Type in 4	Press	$\times$
Type in .875	Press	$=$

and the result of 80.5 will be shown on the display.

Percentages

To work out a percentage of a number you need to remember that the decimal equivalent of a percentage is obtained by dividing the percentage by 100. So 10% becomes 0.10, 13.5% becomes 0.135, and so on. You can then use the decimal equivalent to carry out a multiplication and hence obtain the result. For example, to find 16.5% of 356 785, you would

Type 16.5	Press	$/$
Type in 100	Press	$=$
	Press	$\times$
Type in 356 785	Press	$=$

and the result of 58 869.525 would be displayed.

Powers

The next two keys you will need are

$\sqrt{}$	Square root
x^2	Square

Their meaning is fairly obvious, as long as you remember the BEDMAS rule. For example, evaluate (a) $\sqrt{4} \times 9$ and (b) $\sqrt{(4 \times 9)}$. For (a) you would

Type 4	Press	$\sqrt{}$
	Press	$\times$
Type in 9	Press	$=$

And the result of 18 would be displayed.
   For (b) you would

Type 4	Press	$\times$
Type in 9	Press	$=$
	Press	$\sqrt{}$

and the result of 6 would be displayed.
   Note that we have used the BEDMAS rule here to evaluate the brackets first, and then work out the square root of the answer.
   The next key we need is

Sign change

$+/-$ Sign change

This key will change the sign of whatever is shown on the display, and is useful when you are working out a fairly complicated sum. For example, evaluate $345 - 4 \times 23.986$:

Type 4	Press	$\times$
Type in 23.986	Press	$=$
	Press	$+/-$
	Press	$+$
Type in 345	Press	$=$

and the result of 249.056 would be displayed.

Reciprocal

The last of the standard keys which will be of use to you is

$1/x$ Reciprocal

This will work out 1 divided by whatever is shown on the display and is useful when you are working out division problems. For example evaluate $7534/(4.987 \times 58.9)$:

Type 4.987	Press	$\times$
Type in 58.9	Press	$=$
	Press	$1/x$
	Press	$\times$
Type in 7534	Press	$=$

and the result of 25.649 030 433 29 would be displayed.

Note that we have shown the answer to 11 decimal places. You would not normally quote an answer to this many decimal places; you would probably quote the answer as 25.649.

Finally, if your calculator has a memory, then you can store an intermediate result there whilst you work out something else, and then recall the value in the memory to complete the sum. The buttons you will find on your calculator are probably marked as

M+		Addition to memory
MR		Memory recall
MC	or	*	Memory clear

For example, evaluate

$$\frac{42.47 \times 85.38}{347.8 \times 1.3047}$$

Type in 42.47   Press $\boxed{\times}$

Type in 85.38   Press $\boxed{=}$

                 Press $\boxed{\text{M+}}$

                 Press $\boxed{\text{CE}}$ For clear display

Type in 347.8   Press $\boxed{\times}$

Type in 1.3047   Press $\boxed{=}$

                 Press $\boxed{1/x}$

                 Press $\boxed{\times}$

                 Press $\boxed{\text{MR}}$

                 Press $\boxed{=}$

and the result of 7.990 945 549 934 would be displayed.

You are now ready to go on to the next section, however, you might want to check your understanding of using calculators by working through the Re-tests in Section Six.

Section 3

Algebra

Self-tests on algebra

Self-test

Re-test

Text & Examples

Worked answers

Exercises

What do you already know?

Here we have a series of questions to help you focus on those parts of Section Three which you will need for a good basic understanding of algebra. These two self-tests are about different aspects of algebra, and you may find that you already know many of the ideas involved. When you have tried to answer the questions, check your answers and then make a note of those topics you will need to concentrate on.

Self-test 7

1 You are told that there are currently 0.70 euros to £1. If y is the number of euros and $\times$ is the number of pounds, using an equation, show

(a) how you could convert pounds to euros

(b) how you could convert euros to pounds.

2 It is known that the exchange rate between the euro and the pound can vary. Produce a relationship that will show how to convert pounds to euros given a variable exchange rate (r).

By substitution, answer questions 3 to 5.

3 If $m = 4$, then $3m =$

4 If $a = 1.1$, then $4a + 2 =$

5 If $x = 9$ and $y = 5$, then $2x + 3y + 4y^2 + 11 =$

6 Calculate 15 add 5 subtract −10 add −3

Simplify the expressions given in questions 7 to 12.

7 $a + a + a$

8 $a \times a \times a$

9 $a \times a \times a + a + a$

10 $4x^2 + 2y + x^2 + 2y + 0.5$

11 $\dfrac{x \times x \times x \times x \times x}{x} + y$

12 a^2bb^{-1}

Self-test 8

Simplify the following expressions:

1 $4a^3 + 2a^3 - a^3$

2 $8x^2 + 2x^2 - 2x + 5$

3 $3x^{-2} \times 5x^{-3}$

4 $4x^{-3} \div 8x^{-2}$

5 $\left(a^{-4}\right)^3$

6 $\sqrt[3]{27b^3}$

7 $\left(\sqrt[4]{x^{-16}}\right)^2$

8 $x + (12 + 3) \times y + x$

9 $4(2a - 3b) - 2(2b - c)$

10 $2 + 4 \times 4x - (2x \times x - x + 1)$

Self-test 7 answers

1 (a) to convert pounds to euros:
$y = 0.7x$

(b) to convert euros to pounds
$x = \frac{1}{0.7}y$ or $x = 1.429y$

(you will need to decide what level of rounding to use – see Chapter 3: Decimals and rounding)

2 to convert pounds to euros (given rate of exchange r): $y = r \times x$

3 12

4 6.4

5 144

6 27

7 $3a$

8 a^3

9 $a^3 + 2a$

10 $4y + 5x^2 + 0.5$

11 $x^3 + y$

12 a^2

Self-test 7 is designed to check whether you have the range of skills to work confidently with basic algebra. Questions 1 and 2 are concerned with the construction of equations. Here you are required to use a mix of numbers and symbols to describe a statement, situation or problem. You will need to be able to distinguish between a coefficient, variable and constant. Questions 3 to 5 require substitution of known values into equations. As with numbers, you need to be aware of the 'order of operations' (see Chapter 1: A Rule – BEDMAS) and *remember* always to multiply before you add. Question 6 is a reminder that a minus minus becomes a plus. Questions 7 to 12 are concerned with the basic rules of algebra. Again we would hope that you have got them all right. If you got questions 1 and 2 wrong, then you will need to read the first part of Chapter 7 concerned with forming equations. If you had some difficulty with the remaining questions then you will need to revise some of the basic ways of working with algebra and you should now work through Chapter 7.

Self-test 8 answers

1 $5a^3$

2 $10x^2 - 2x + 5$

3 $15x^{-5}$

4 $0.5x^{-1}$ or $\frac{1}{2x}$

5 a^{-12}

6 $3b$ (note $3 \times 3 \times 3 = 27$)

7 x^{-8}

8 $15y + 2x$

9 $8a - 16b + 2c$

10 $-2x^2 + 17x + 1$

Self-test 8 is designed to test whether you can work with a more difficult algebra that involves a number of different terms and a number of different operations. Questions 1 and 2 are concerned with addition and subtraction, and the importance of using such operations only with like terms (e.g. you add or subtract separately all the squared or cubed terms for example). If you have a problem with this sort of question you should consider looking, again if necessary, at Chapter 7. Questions 3 and 4 are concerned with the multiplication and division of terms which requires the correct addition (when multiplying) and correct subtraction (when dividing) of powers. Question 5 involves raising a power term by a power. Questions 6 and 7 require a knowledge of fractional powers. Even if you are familiar with this more advanced use of powers, you may still find it useful to look at the general expressions given in Chapter 8 for powers of powers and fractional powers. Questions 8 to 10 are concerned with the use of brackets and the order of operations (hopefully you will remember BEDMAS – see Chapter 1).

CHAPTER 7

The basics

Basic algebra is a part of everyday life but we do not always think about it in that way. Whenever you manipulate quantities you are doing more than just mental gymnastics, you are doing algebra.

Algebra

Working out the cost of a basket of goods or the time to travel a given distance may just involve the use of a few basic sums but you can represent and often simplify by using the rules of algebra. If the cost of travelling one mile is 40 pence, then the cost of travelling 5 miles is 200 pence or £2.00. If we were to make three journeys of this kind, then the cost of travelling would be £6.00. If we also need to allow £1.20 for the cost of parking each time we travelled then the cost of one 5 mile journey would become £3.20 and the cost of three journeys of this kind would become £9.60.

To represent the cost of travelling in algebraic terms we need to use **letters** or **symbols**. If m is the number of miles travelled and the cost of travel is 40 pence per mile,

$$\text{then the cost of travelling (in £s)} = 0.40 \times m$$

If the number of miles is 5, then by letting $m = 5$, we find the cost to be £2.00. If the number of miles is 17, then by letting $m = 17$ (we say **by substituting** m equal to 17), we find the cost to be £6.80.

To allow for the fixed cost of £1.20 for parking, we would just add this constant amount:

$$\text{the cost of travel (in £s) becomes} = 0.40 \times m + 1.20$$

If 6 miles were travelled the cost (including parking) becomes $0.40 \times 6 + 1.20$ which equals £3.60. It is important to note that we have multiplied first then added the fixed parking fee of £1.20 (see Chapter 1: A Rule – BEDMAS).

Spreadsheets are now used extensively to describe, model and help solve a wide range of problems. You will find that a good understanding of algebra will help you construct spreadsheets (see Chapter 14).

An **algebraic expression** is a set of letters, symbols and numbers linked by the arithmetic operators $+$, $-$, $\times$ and $\div$. The algebraic expression used in the above example would typically be written:

$$0.40m + 1.20$$

The multiplication sign is not usually included when the need to multiply is obvious:

2a is written rather than $2 \times a$

and

$2x + 3y$ is written rather than $2 \times x + 3 \times y$

The algebra expression $0.40m + 1.20$ has two terms, $0.40m$ and $+ 1.20$.

The number 0.40 is referred to as the **coefficient**,
m as the **variable**
and $+1.20$ (or just 1.20) as the **constant**.

We can describe these more precisely as:

a **coefficient** (within the context of an equation) is placed just before
a variable and will be used for multiplication (e.g. 40 pence per mile),
a **variable** is a particular characteristic of interest, such as height,
time, weight, cost and can take the range of values that measurement
allows (e.g. number of miles travelled), and
a **constant** is a characteristic or quantity that does not vary (e.g. the
fixed cost of parking of £1.20).

If the cost of parking were to change from a fixed cost of £1.20 (a constant) to a
charge of 50 pence per hour, then the total cost of parking would become variable
(would depend on time). We could write the new expression as

cost of parking (in £s) $= 0.50t$

where t is the time in hours (a variable)

and 0.50 the coefficient

Algebra uses all the arithmetic operations $(+, -, \times, \div)$ that you are already familiar
with (see Chapter 1) and follows the same rule for the **order of operations** (also in
Chapter 1).

Addition

To avoid the repetitive activity of writing or adding the same number over and over
again, we naturally use multiplication. Just imagine that we had just bought 10
'Beanie Babies' at £4.99 each. Rather than tackle the larger sum

£4.99 + £4.99 + £4.99 + £4.99 + £4.99 + £4.99 + £4.99 + £4.99 +
£4.99 + £4.99

we would probably choose to multiply £4.99 by 10:

£4.99 $\times$ 10 $=$ £49.90

In algebra, the **coefficient** is telling us how many times we are effectively adding:

$a + a + a + a + a + a = 6a$

$$a + a + b + b + b = 2a + 3b$$

If we are asked to add a negative quantity, then the negation remains:

$$a \text{ plus } -b = a - b$$

more usually written with brackets as $a + (-b) = a - b$

$$a + a + a \text{ plus} - a = 2a$$

more usually written as $a + a + a + (-a) = 2a$

Subtraction

The important thing to remember here, is that if you subtract a negative quantity (a minus minus), then the outcome is positive. If someone were to take a debt of £10 away from you, you would actually be better off by £10.
  In algebra, we observe a sign change:

$$a \text{ take away } - b \text{ is } a + b$$

more usually written as $a - (-b) = a + b$

$$a + a + a \text{ take away } - a \text{ is } 4a$$

more usually written as $a + a + a - (-a) = 4a$
  Brackets are used to clarify the negative quantity:

$$14a - (-2a) = 16a$$

$$5a + 4b - (-3a) - 2b = 8a - 2b$$

Multiplication

If a quantity is multiplied by itself over and over again, we can avoid repetitive writing by using a number called an **index** or a **power**.
  In the same way we could write

$$3 \times 3 \times 3 \times 3 = 3^4$$

we can also write

$$a \times a \times a \times a = a^4 \text{ (and would say } a \text{ to the power 4)}$$

Multiplication can also include a mix of terms:

$$3 \times a \times a = 3a^2$$

$$5 \times a \times b \times b \times b = 3ab^3$$

The order of multiplication is not important

$$a \times b = b \times a$$

but

$$b \times a \times a \times b \times 10$$

would typically be written $10a^2b^2$.

It should also be noted that terms such as a^2 and a^3 are treated separately:

$$3 \times a \times a + 2 \times a \times a \times a$$

would be written

$$3a^2 + 2a^3$$

Division

In its simplest form, division can just involve cancelling out equal quantities above and below a dividing line:

$$\frac{4 \times 4 \times 4 \times 4}{4 \times 4} = 4 \times 4$$

$$\frac{a \times a \times a \times a}{a \times a} = a \times a = a^2$$

Expressions can also include a mix of terms:

$$\frac{8 \times a \times a \times b \times b}{2 \times a \times b \times b} = 4a$$

A negative power or index indicates a divisor:

$$a^{-1} = \frac{1}{a}$$

$$a^2 b^{-1} = \frac{a^2}{b}$$

Exercises

1 The cost of a new parking scheme has two elements: a fixed charge of 70 pence and a charge of 40 pence per hour. If c is used to represent the total cost of parking in £s and t is the time in hours, then

(a) express the relationship between total cost of parking and t by an equation,

(b) use the equation from part (a) to calculate the total cost of parking for a stay of 7 hours.

2 To convert temperature measured in degrees celsius (centigrade) to degrees fahrenheit, you are told to multiply degrees centigrade by 1.8 and add 32.

(a) Using c to represent degrees centigrade and f to represent degrees fahrenheit, produce an equation to show the required conversion.

(b) Use your equation from part (a) to convert 12 degrees centigrade to fahrenheit.

By substitution, answer questions 3 to 5.

3 If $x = 5.5$, then $10x =$

4 If $a = 2.4$ and $b = 3.2$, then $a + 2b =$

5 If $a = 4$ and $b = 3$, then $2a + 4b^2 =$

6 Calculate 9 add −1 subtract 5 add 2 subtract −3

Simplify the expressions given in questions 7 to 12.

7 $a + a \times a \times a$

8 $3a^2 + 2b + 2a + 2a^2$

9 $x + y \times x \times x + 2x^2 - 4x^2 - x$

10 $8a^2 - 3a \times a + b \times a - b^3$

11 $\dfrac{a \times a \times b \times a \times c \times c}{a \times c} + 3a^2bc$

12 $x^3 y^2 x^{-1} z y^{-1}$

Worked answers

1 (a) $c = 0.70 + 0.40t$
where 0.70 is the fixed cost element   *(expressed in £s) which does not change with time,*

and +0.40 is the variable cost element   *(expressed in £s) which increases with time (t)*

(b) Let $t = 7$ *to substitute*
$$c = 0.70 + 0.40 \times 7$$
$$= 0.70 + 2.80$$
$$= £3.50$$

2 (a) $f = 1.8c + 32$ *1.8 is the coefficient and 32 is the constant*
(b) Let $c = 12$ *to substitute*
$$f = 1.8 \times 12 + 32$$
$$= 21.6 + 32$$
$$= 53.6$$

3 If $x = 5.5$, then $10x = 10 \times 5.5 = 55$

4 If $a = 2.4$ and $b = 3.2$, then $a + 2b = 2.4 + 2 \times 3.2$
$$= 2.4 + 6.4 = 8.8$$

5 If $a = 4$ and $b = 3$, then $2a + 4b^2 = 2 \times 4 + 4 \times 3 \times 3$
$$= 8 + 36 = 44$$

6 $9 + (-1) - 5 + 2 - (-3) = 9 - 1 - 5 + 2 + 3 = 8$

7 $a + a \times a \times a = a + a^3$ *a trick question*

8 $3a^2 + 2b + 2a + 2a^2 = 2a + 5a^2 + 2b$ *collecting terms*

9 $x + y \times x \times x + 2x^2 - 4x^2 - x$
$$= x - x + yx^2 + 2x^2 - 4x^2 = -2x^2 + yx^2$$

10 $8a^2 - 3a \times a + b \times a - b^3$
$$= 8a^2 - 3a^2 + ab - b^3 = 5a^2 + ab - b^3$$

11 $\dfrac{a \times a \times b \times a \times c \times c}{a \times c} + 3a^2bc = a^2bc + 3a^2bc = 4a^2bc$

12 $x^3 y^2 x^{-1} z y^{-1} = x^2 yz$ *summing powers of like terms*

You are now ready to go on to Chapter 8, or you might want to check your understanding of algebra basics by working through the Re-tests in Section Six.

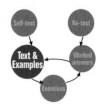

CHAPTER 8

Working with powers, brackets and order

To manipulate an expression like $4a^3 + 2a^3 - a^3$ or $\frac{-b \pm \sqrt{b^2 - 4ac}}{2a}$ may not seem like part of everyday life, but your course may expect you to do it. You may have already seen that powers (see Chapter 5) provide a useful sort of shorthand. If you were to say '2 to the power 3' it would be known that you meant 2 multiplied by itself 3 times:

$$2^3 = 2 \times 2 \times 2 = 8$$

In the same way 'a to the power 3' would be written:

$$a^3 = a \times a \times a$$

The power term, in this case 3, is also referred to as an **index**.

Adding and subtracting

We can further simplify expressions by collecting together terms with the same power:

$$x^2 + 2x^2 + 4x^2 = 7x^2$$

but expressions with terms of different powers would treat the powers separately, so

$$3 \times a \times a + 2 \times a \times a \times a$$

would be written:

$$3a^2 + 2a^3$$

(see Chapter 7).

Only terms with the same letter, symbol or combination of letters and symbols, and the same power can be added or subtracted, for example:

$$2x^2 + 4y + x^2 - 3y$$

$$= (2x^2 + x^2) + (4y - 3y)$$

$$= 3x^2 + y$$

Multiplication   We have already shown (see Chapter 5) that when a number raised to a power is multiplied by the same number raised to a power, then we can **add the powers**:

$$2^3 \times 2^2 = 2^{3+2} = 2^5 = 32$$

(or 8 × 4 = 32)

This can be expressed as a **rule**:

$$a^m \times a^n = a^{m+n}$$

Collecting together multiplicative terms does allow us to simplify many complex expressions, for example:

$$a^5 \times a^3 = a^8$$

If the expression includes a number of different letters and symbols, we may decide to collect together these terms first:

$$a^2 b^2 c \times ab^2 c$$

which can be written as:

$$\left(a \times a^2\right) \times \left(b^2 \times b^2\right) \times c \times c$$

$$a^3 \times b^4 \times c^2 \text{ or perhaps more simply } a^3 b^4 c^2$$

Division   We have also shown (see Chapter 5) that division involves the **subtraction of powers**:

$$3^4 \div 3^2 = 3^{4-2} = 3^2 = 9$$

$$\left(\text{or } \frac{3 \times 3 \times 3 \times 3}{3 \times 3} = 3 \times 3 = 9\right)$$

This can also be expressed as a **rule**:

$$a^m \div a^n = a^{m-n}$$

Collecting together terms again allows simplification, for example:

$$a^7 \div a^5 = a^{7-5} = a^2$$

and

$$12x^5 \div 3x^3$$
$$= \frac{12}{3} \times \left(\frac{x^5}{x^3} \right)$$
$$= 4 \times x^{5-3}$$
$$= 4x^2$$

A **negative power** indicates a divisor:

$$a^{-n} = \frac{1}{a^n}$$

so

$$a^6 \div a^2 = a^{6-2} = a^4$$

which is no different from

$$a^6 \times a^{-2} = a^{6+(-2)} = a^4$$

or

$$\frac{a \times a \times a \times a \times a \times a}{a \times a} = a^4$$

Power of powers

In the same way that

$$\left(3^2 \right)^4 = 3^{2\times4} = 3^8 = 6561$$

requires a multiplication of power terms, in general:

$$\left(a^m \right)^n = a^{m \times n}$$

However, we may need to break down an expression into parts, for example:

$$\left(5y^3 \right)^2$$
$$= 5^{1\times2} \times y^{3\times2}$$
$$= 5^2 \times y^6$$
$$= 25y^6$$

Fractional powers

In this section the ideas of the **square root**, **cube root** and other fractional powers will be developed. Just as a reminder, a **square root** of a number is that number which, when multiplied by itself gives, the original number, and a **cube root** is that number which, when multiplied by itself and then multiplied by itself again, gives the original number.

As $3 \times 3 = 9$, the square root of 9 is equal to 3 or -3 (you should remember that a negative multiplied by a negative gives a positive).

The square root can be written as:

$$\sqrt{9} = \pm 3 \quad \text{or} \quad (9)^{\frac{1}{2}} = \pm 3$$

As $6 \times 6 \times 6 = 216$, the cube root of 216 is equal to 6. In this case the cube root is positive as a negative multiplied by a negative multiplied by a negative would always be negative. The cube root of -216 would equal -6.

The cube root can be written as:

$$\sqrt[3]{216} = 6 \quad \text{or} \quad (216)^{\frac{1}{3}} = 6$$

The notation for a square root $a^{\frac{1}{2}}$ or $\sqrt{a}$ can easily be extended to the more general n^{th} root:

$$a^{\frac{1}{n}} = \sqrt[n]{a}$$

As an example, consider the following:

$$(a \times a) \times (a \times a) \times (a \times a) = \left(a^2\right)^3 = a^6$$

so

$$\sqrt[3]{a^6} = a^2$$

as a^2 multiplied by itself three times gives a^6.

We can use this concept of a fractional power to simplify expressions:

$$\sqrt[3]{a^6} = a^{\frac{6}{3}} = a^2$$

In general, the nth root of a^{m}:

$$\sqrt[n]{a^m} = a^{\frac{m}{n}}$$

Given a more complex expression such as:

$$y = \left(\sqrt[3]{x^{-15}}\right)^4$$

we can simplify using the **rule** given by $\sqrt[n]{a^m} = a^{\frac{m}{n}}$

$$y = \left(x^{\frac{-15}{3}}\right)^4$$

$$= \left(x^{-5}\right)^4$$

and can further simplify using $(a^m)^n = a^{m \times n}$

$$y = x^{-5 \times 4} = x^{-20}$$

or $y = \dfrac{1}{x^{20}}$

Brackets

In this section only the basic use of brackets is considered. Brackets are essentially an instruction that the work within the brackets needs to be completed first:

$$(8+3) \times a = 11a$$

A coefficient attached to the brackets is used to multiply **each** of the parts contained within the brackets:

$$3(a + 2b) = 3a + 6b$$

$$5(x - y + x - 2y) = 5(2x - 3y) = 10x - 15y$$

Order of operations

We have already seen (see Chapter 1) that the order of operations can be summarized as:

B	Brackets
E	Exponentiation
D	Division
M	Multiplication
A	Addition
S	Subtraction

For example, if we take the expression

$$3(2x + x + 1) + 6x + 7 \times 2x - 4 - 2x$$

we would work out the value within the bracket:

$$3(3x + 1) + 6x + 72x - 4 - 2x$$

we would then multiply the contents of the brackets by 3 and multiply $2x$ by 7:

$$9x + 3 + 6x + 14x - 4 - 2x$$

we would then do the additions:

$$29x + 3 - 4 - 2x$$

and finally the subtractions:

$$27x - 1$$

Exercises

1 $2a^2 + a^2 + a - a \times a$

2 $4b^4 - b \times b^3 + 4b - b$

3 $2y^2 \times 4y^{-3} \times y^{-1}$

4 $8x^{-2} \div 2x^{-4}$

5 $\left(4b^{-5}\right)^3$

6 $\sqrt[4]{16x^{-8}}$

7 $\left(\sqrt[3]{x^3}\right)^2$

8 $(x + y) \times 5 + 2x$

9 $3(2x - y) - 2x \times 4$

10 $4x + 2 \times 4x - 5(2x \times 4 - 3x)$

Worked answers

1 $2a^2 + a$ *collect together the a^2 and a terms separately*

2 $3b^4 + 3b$ *same again*

3 $8y^{-2}$ or $\dfrac{8}{y^2}$ *power terms can be added*

4 $4x^2$ *power terms can be subtracted*

5 $64b^{-15}$ *power terms are multiplied*

6 $2x^{-2}$ *'fractional powers'*

7 x^2

Questions 8 to 10 are all concerned with the use of brackets and the order of operations.

8 $5x + 5y + 2x = 7x + 5y$

9 $6x - 3y - 8x = -2x - 3y$

10 $4x + 8x - 5 \times 5x = -13x$ *bracket and multiply both done together*

You are now ready to go on to the next section, but if you have not already done so, it would be a good idea to go to Section Six to check your understanding of algebra.

Section 4

Equations and Graphs

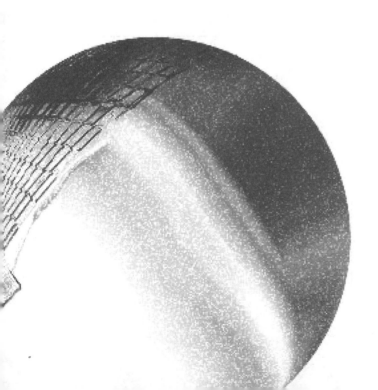

Self-tests on equations and graphs

What do you already know?

Here we have a series of questions to help you focus on those parts of Section Four which you will need for a good understanding of equations and graphs. These five self-tests cover a range of subjects, and you may find that you already know many of the ideas involved. When you have tried to answer the questions, check your answers and then make a note of those topics you will need to concentrate on.

Self-test 9

1 To produce a component involves a company in two types of cost. One is a fixed cost of £5,000 which covers items such as rent, and the other is a cost of £1.50 per component which covers items like labour and materials. If c is used to represent total cost and x is used to represent the number of components, express the total cost to the company by means of an equation.

2 A telephone answering service operates from two call centres, one in the Northern region and one in the Southern region. If a total of 6,600 calls where received, and the call centre in the Northern region answered twice as many calls as the call centre in the Southern region, express the problem by means of an equation and determine the number of calls answered by each centre.

3 If travel cost is represented by the equation $y = 0.42x + 7.50$, where y is the total travel cost in £s and x is the number of miles travelled, determine total travel costs if the number of miles travelled is 135.

4 Using the equation given in question 3, determine the number of miles travelled to incur a total travel cost of £39.84.

5 Determine the value of s given that $s = 21 + 6t$ and that $t = 9$.

Solve the following equations:

6 $8x = 2x + 42$

7 $3a = 2(a - 2) + 6$

8 $4b - 3(2 - b) = 5(b + 2)$

Self-test 10

Draw appropriate axes and (on the same graph) plot the points A, B, C, D and E as defined by the co-ordinates given in questions 1 to 5:

1 A: (2, 3)

2 B: (−3,−2)

3 C: (−4, 5)

4 D: (0,−1)

5 E: (2,−4)

6 Draw a graph showing how sales have changed over time using the following data:

Year	Annual sales (in units)
1995	1,800
1996	1,820
1997	1,815
1998	1,850
1999	1,860

7 The following table shows the number of warranty claims against a particular product on a quarterly basis.

Year	Quarter 1	Quarter 2	Quarter 3	Quarter 4
1997	77	96	104	88
1998	74	89	99	80
1999	65	80	91	

Draw a time series graph (plot the number of warranty claims against time).

8 A company is interested in the relationship between the processing time and the length of life of a component. The following information has been provided from a sample of 10 components.

Processing time (in seconds)	Length of life (in days)
2	4
20	57
4	10
6	31
9	29
16	45
13	34
6	20
18	30
19	43

Draw a graph to show the relationship between the length of life of a component and processing time.

Self-test 11

Draw appropriate axes and on the same graph plot the following:

1 $y = 5 + x$ between $x = -4$ and $x = 3$

2 $y = 1$ between $x = -4$ and $x = 4$

3 $x = 3$ between $y = -2$ and $y = 8$

4 Using appropriate axes, plot the following points

x	y
−3	−10
0	2
1	6
4	18
5	22

5 On the graph you have just drawn (question 4), join the points with a straight line. Now determine the linear equation that describes this line. (Hint: it should take the form $y = a + bx$.)

Self-test 12

1 Given the following equations

$$y = 2x + 1$$
and $$y = \frac{1}{2}x + 5.5$$

(a) plot both on the same graph and identify the point of intersection (the point where the lines cross)

(b) solve as simultaneous equations.

Solve each of the following pairs of simultaneous equations using appropriate algebra:

2 $3x + y = 14$
$5x - y = 10$

3 $2x + 3y = 10$
$x - 4y = -17$

4 $2x + 3y = 15$
$4x + y = 11$

5 $2x + 3y = 16$
$3x - y = 13$

Self-test 13

1 Draw the graph of $y = x^2 - 5x + 4$ from $x = 0$ to $x = 6$ and identify the roots.

Using the method of factorization, find the roots of the following quadratic equations:

2 $x^2 - 5x + 4 = 0$

3 $x^2 - 3x - 4 = 0$

4 $x^2 - 36 = 0$

5 $x^2 - x - 20 = 0$

Using the equation $x = \dfrac{-b \pm \sqrt{\left(b^2 - 4ac\right)}}{2a}$ solve the following quadratic equations

6 $x^2 - 5x + 4.56 = 0$

7 $x^2 + 2.2x - 9.68 = 0$

Self-test 9 answers

1 $c = 1.50x + 5000$

2 $2x + x = 6600$
$x = 2200$

The Northern region answers 4,400 calls and the Southern region answers 2,200 calls.

3 $y = 0.42 \times 135 + 7.50$
$= £64.20$

4 $39.84 = 0.42x + 7.50$
$x = 77$ miles

5 75

6 $x = 7$

7 $a = 2$

8 $b = 8$

The first two questions in this self-test are concerned with the construction of equations. In particular, you need to be familiar with the representation of an unknown quantity (e.g. the number of components to be produced or share of calls) with a symbol or letter. Questions 3 to 5 require the substitution of a value to find a solution. Questions 6 to 8 require the solution of simple equations. What you need to remember when solving equations is **whatever you do to one side of an equation you will also need to do to the other** (to maintain the equality). If you had difficulty with any of these questions, read Chapter 9.

Self-test 10 answers

The graph required for questions 1 to 5 is shown below:

Figure ST10.1

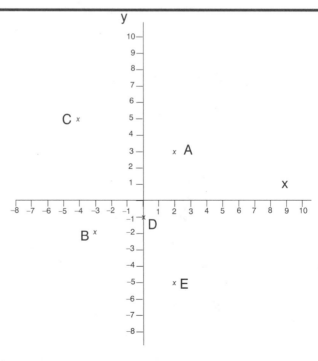

6

Figure ST10.2

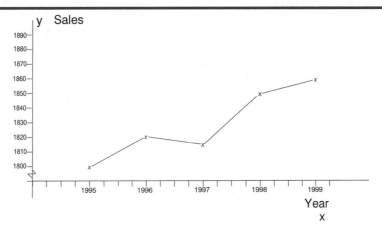

7

Figure ST10.3

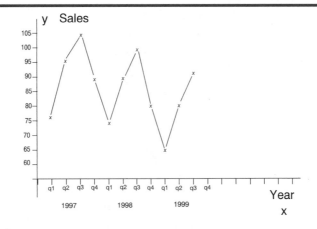

8

Figure ST10.4

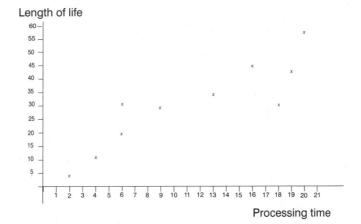

The first five questions are concerned with the correct use of co-ordinates; within the brackets the x value is given first and the y value second. You should particularly note that negative x values lie to the left of the origin (0,0) on the horizontal line and negative y values lie below the origin on the vertical line. Questions 6 and 7 are concerned with plotting values against time. You should note how the x axis is scaled in each case. Quarterly data as shown in the answer to question 7 provide an opportunity to examine both the general trend over time (downwards in this case) and the predictable quarterly differences (quarter 3 is typically higher than the other quarters for example). Question 8 examines the relationship between two variables; in this case the length of life of a component (in days) and processing time (in seconds). Processing time is plotted against the x axis because this is the one we can control and length of life is plotted against the y axis because that is the one we would like to be able to predict. The concept of an independent variable and a dependent variable is explained within the text. If you had difficulty with any of these questions, read Chapter 10.

Self-test 11 answers

The graph required for questions 1 to 3 is shown below:

Figure ST11.1

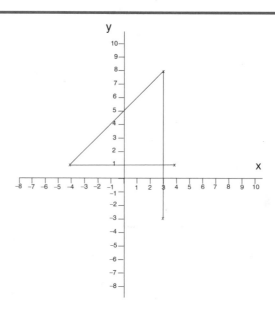

Did you get the figure 4?

The graph below shows the plotting of points from question 4 and the determination of the equation $y = 2 + 4x$

Questions 1 to 3 are all concerned with representing linear equations as straight lines on graphs. To plot the first equation, $y = 5 + x$, you need to establish two points to join with a ruler line. In this case, it is probably most convenient to use the more extreme values of $\times$ (given the range $x = -4$ to $x = 3$). If $x = -4$ then $y = 1$, and if $x = 3$, then $y = 8$; joining these two points gives the first line. The second equation, $y = 1$, produces a horizontal line as the value is not influenced by x. The third equation, $x = 3$, produces a vertical line as the value is not influenced by y.

Figure ST11.2

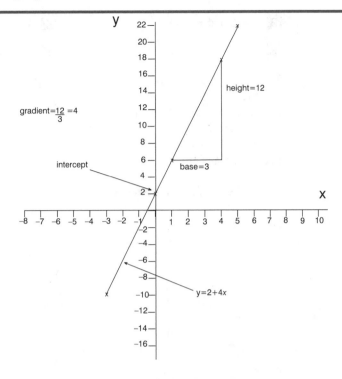

Questions 4 and 5 require the plotting of a set of points, a straight line being drawn through them, the determination of the intercept (the point where the straight line cuts the y axis) and the gradient (the increase in y given a unit increase in x). If you had difficulty with any of these questions, read Chapter 11.

Self-test 12 answers

1 (a)
   (b) $x = 3$; $y = 7$

Figure ST12.1

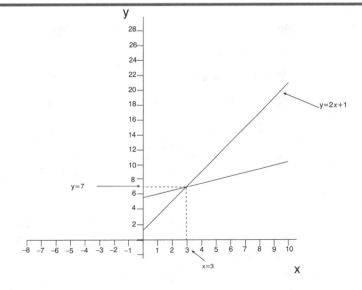

2 $x = 3$ and $y = 5$

4 $x = 1.8$ and $y = 3.8$

3 $x = -1$ and $y = 4$

5 $x = 5$ and $y = 2$

The first question shows how two linear equations can be represented graphically and the point where the lines cross corresponds to their solution as simultaneous equations. Clearly, if the lines were parallel no cross-over would exist and no solution could be found by trying to solve as simultaneous equations. Questions 2 to 5 give you the opportunity to try solving a few simultaneous equations. If you had difficulty with any of these questions, read Chapter 12.

Self-test 13 answers

1

Figure ST13.1

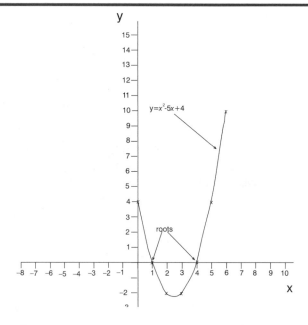

$y = x^2 - 5x + 4$

roots

2 The roots are $x = 1$ and $x = 4$

3 The roots are $x = -1$ and $x = 4$

4 The roots are $x = -6$ and $x = 6$

5 The roots are $x = -4$ and $x = 5$

6 The roots of the quadratic are $x = 3.8$ or $x = 1.2$

7 The roots of the quadratic are $x = 2.2$ or $x = -4.4$

Question 1 provides a check that you can plot a quadratic equation. Given that a curve is produced you can no longer use two points and join with a ruler (like linear functions) but need sufficient points clearly to draw the curve shape. Questions 2 to 5 are checking whether you can still remember how to factorize. It may be that a little practice is all you need but you may also need to consider whether you need the systematic approach outlined in the text that follows. The last two questions are intended as a reminder that the formula can be used, and that if the numbers are more difficult, the formula may provide the only easy means of solution. If you had difficulty with any of these questions, read Chapter 13.

CHAPTER 9

The basics

An equation is nothing more than a mathematical statement that two things, expressions or quantities, are equal. How the cost of travel changes with distance or how to convert from inches to centimetres can both be represented by equations:

$$c = 0.40m$$

where c is the total cost of travel (£s),
    0.40 is the cost of travelling one mile in £s, and
    m is the number of miles travelled

$$c = 2.54i$$

where c is the length in centimetres, and
    i is the length in inches.

Whatever the equation represents you need to be sure what the symbols mean (we have used c in both equations given above but their meaning is very different) and what the units of are (e.g. miles, £s, $s).

The solution of a problem using the concept of equality (equations), will usually involve one or more of the following stages: **forming**, **simplification**, **the collection of terms**, **the substitution of values**.

We will also need to remember a fundamental principle that:

whatever you do to one side of an equation you must also do to the other.

Forming an equation

Having established the nature of the equality, unknown quantities (variables), are represented by symbols or letters such as x, y, a or b.

If the cost of ordering material of a particular kind, for example, is £2.25 per metre plus £5.00 postage and packing, we can represent as an equation:

$$c = 2.25x + 5.00$$

where c is total cost in £s,
    x is the unknown number of metres being ordered, and
    5.00 is the known fixed cost in £s of postage and packing.

Consider now an agreement made by Mrs Smith and Mr Singh that the profit from a certain business activity should be shared in proportion to the hours they have worked. If Mr Singh worked three times as many hours as Mrs Smith and the profit was £1,200, how much should each get?

If Mrs Smith is to get x, then Mr Singh should get $3x$. As all the profits go to Mrs Smith and Mr Singh, we can write

$$x + 3x = 1200$$
$$4x = 1200$$
$$x = 300$$

In this case, Mrs Smith should get £300 (x) and Mr Singh should get £900 ($3x$). (*This is similar to the ratio calculations in Chapter 4.*)

Simplification

Equations can be presented in a variety of ways. Essentially, we want to bring all the like terms together (all the xs, all the ys, all the as and all the a^2s). **Do remember that a and a^2 are treated separately.**

Given $4a(2a - 3) - 8a^2 + 14a = 26$

we work out the brackets first (remembering BEDMAS) which gives

$$8a^2 - 12a - 8a^2 + 14a = 26$$

We can then collect the terms and further simplify:

$$2a = 26$$
$$a = 13$$

Collecting terms

We have already seen how bringing terms together can simplify an expression (e.g. $-12a + 14a$ can be combined to become just $2a$). The terms can be on either side of the equals sign, and we must remember

whatever we do to one side of an equation we must do to the other.

If $8a = 6a + 24$

we can subtract $6a$ from both sides

$$8a - 6a = 6a - 6a + 24$$

to get $2a = 24$
and dividing both sides by 2 gives

$$a = 12$$

Suppose now that we need to collect together the terms of the following equation:

$$4x + 5y = -3x + 11y + 13$$

we can add $3x$ to both sides

$$4x + 5y + 3x = -3x + 11y + 13 + 3x$$
$$7x + 5y = 11y + 13$$

we can then subtract $11y$ (with experience we tend to just do it rather than write it out again showing the particular addition or subtraction) from both sides

$$7x - 6y = 13$$

and then finally subtract 13 from both sides

$$7x - 6y - 13 = 0$$

We can also be asked to express one variable in terms of another. Given an expression like $20x + 5y = 30$, we could be asked to express y in terms of x.
    Subtract $20x$ from each side:

$$20x + 5y - 20x = 30 - 20x$$
$$5y = -20x + 30$$

We can then divide both sides by 5 to get:

$$y = -4x + 6$$

If the cost of ordering material of a particular kind is given by

$$c = 2.25x + 5.00$$

where c is total cost in £s,
            x is the unknown number of metres being ordered, and
            5.00 is the known fixed cost in £s of postage and packing

we can easily calculate the costs of an order by substitution.
    For example, if 10 metres had been ordered, we would let x = 10:

$$c = 2.25 \times 10 + 5.00$$
$$= 22.50 + 5.00$$
$$= £27.50$$

Similarly, if 17 metres had been ordered, we would let x = 17 and

$$c = 2.25 \times 17 + 5.00$$
$$= 38.25 + 5.00$$
$$= £43.25$$

When working with equations, if we are given values for all the other variables, we can always calculate the value for the remaining variable. If, for example, we were told that the cost of the order was £38.75, we could then calculate how many metres had been ordered:

$$38.75 = 2.25x + 5.00$$

We can then take 5 from both sides:

$$38.75 - 5 = 2.25x + 5.00 - 5$$
$$33.75 = 2.25x$$

If we then divide by 2.25:

$$15 = x$$

or more conveniently $x = 15$ metres.

Another example: if we are given the expression

$$s = 7u + 8v$$

and told that $s = 53$ and $v = 4$, we could solve in terms of u:

$$53 = 7u + 8 \times 4$$
$$53 = 7u + 32$$
$$21 = 7u$$
$$u = 3$$

Exercises

1 A company sends two types of leaflet through the post. The smaller leaflet costs 30 pence to process and post, and the larger leaflet 38 pence to process and post. The equipment hire costs are £280 per week. Construct an equation to show the total cost (in £s) of the postal operation using c to represent total postal cost in £s, x to represent the number of smaller leaflets sent and y to represent the number of larger leaflets sent.

2 A new type of cereal is being produced that will be sold in packs of 500 gms. The cereal will be made from a mix of bran flakes and fruit. If the content of bran flakes is intended to weigh 3 times that of the fruit content, how much of each ingredient is required. If bran flakes cost the manufacturer 18 pence per 100gms and fruit 32 pence per 100 gms, what will be the cost of producing each pack of cereal?

3 A company has decided to use the following equation to estimate the cost of deliveries:

$$c = 1.20x + 8.90t + 30$$

where c is the total cost in £s,
x is the number of miles, and
t is the time taken in hours.

Use the equation to estimate the travel cost of a 21 mile journey that took 2½ hours.

4 Using the equation given in question 3, determine the travel time for a journey of 15 miles that had an estimated cost of £61.35.

5 Determine the value of u given that $u = 3 + 6y + 4y^2$ and $y = 7$.

Solve the following equations:

6 $2 - 8x = 12 - 4x$

7 $16a = 4(2a - 2) - 16$

8 $4y - 5 \times y + y \times 6 = -5(y + 2)$

Worked answers

1 $c = 0.30x + 0.38y + 280$ In this example, there are two sources of variable cost (the posting of the smaller and larger leaflet) and the fixed cost of £280.

Where units are all given in £s.

2 $500 = x + 3x$ *where x is the weight of fruit and 3x is the weight of bran*

$x = 125$ gms.   *Solving the equation*
Therefore the cereal requires 125 gms of fruit (x) and 375 gms of bran flakes ($3x$).

The cost of fruit is 40 pence   *Now we can work out the costs*
($125 \times 32/100$) and the cost of bran flakes 67.5 pence ($375 \times 18/100$)

3 $c = 1.20 \times 21 + 8.90 \times 2.5 + 30$ *substituting values of x and t*
$ = 25.20 + 22.25 + 30$
$ = £77.45$

4 $61.35 = 1.20 \times 15 + 8.90\,t + 30$
substitution
$61.35 = 18 + 8.90t + 30$ *rearranging to find t*
$13.35 = 8.90\,t$
$ t = 1.5\ (1\frac{1}{2}\ \text{hours})$

5 $u = 3 + 6 \times 7 + 4 \times (7)^2$ *substitution*
$ = 3 + 42 + 196$
$ = 241$

6 $-4x = 10$ *rearranging and collecting terms*
$x = -2.5$

7 $16a = 8a - 8 - 16$
$8a = -24$
$a = -3$

8 $4y - 5y + 6y = -5y - 10$
$10y = -10$
$y = -1$

You are now ready to go on to Chapter 10, or you might want to check your understanding of the basics of equations by working through the Re-tests in Section Six.

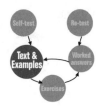

CHAPTER 10

Graphs

Graphs are essentially used in three ways. They can be used to show changes over time (referred to as a **time series** or **time series data**), to explore the relationship between factors (**the relationship between variables**), or to represent known functional relationships (**equations**).

In this chapter, we shall first consider the definition of **axes**, then the use of **coordinates**, the plotting of values against time, and finally plotting one variable against another. The plotting of equations will be covered in the remaining chapters of this section.

 The axes

A graph is constructed by drawing two lines at right angles to each other. A scale is added using 'ruler' measurement. The **vertical axis** is labelled y and the **horizontal axis** is labelled x. This distinction is regarded as particularly important as y is seen as being dependent on x or predictable from x. The graph shown in Figure 10.1 is typical. You will see a number of examples where y and x are referred to more specifically in terms of the subject and units of measurement (e.g. weight in gms, process time in seconds).

Figure 10.1

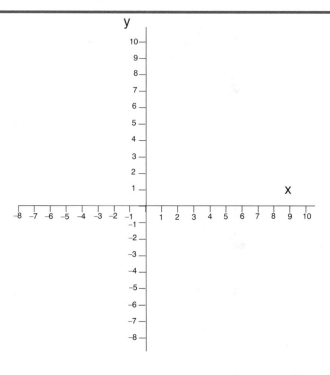

The horizontal line is referred to as the *x*-axis and the vertical line is referred to as the *y*-axis. The point at which the lines cross is called the **point of intersection** or **origin**. Any point on the graph can be identified by reference to the horizontal position (the *x* value) and the vertical position (the *y* value) – *always in that order*. In general, the point of reference is given in the form (a, b) where a is the *x* value and b is the *y* value.

The origin, then is given by (0,0). A point defined by *x* = 3 and *y* = 5, is referred to as having the co-ordinates (3,5) and this is shown as point A on Figure 10.2. A point with the co-ordinates (−4,2) is shown as point B.

Figure 10.2

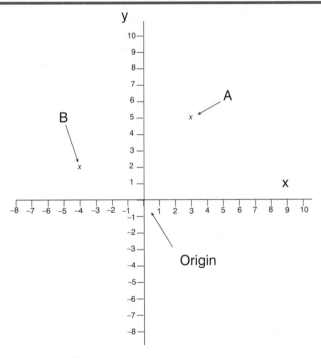

Plotting a time series

One particularly useful graph is the plotting of data against time. Change over time can be of interest for many reasons. Governments may wish to monitor prices, or employment or the balance of payments over time. The government, and society in general, may be interested in other trends over time, such as the number of smokers, changes in diet or the number of teenage pregnancies. Companies will record sales and other business information over time. Individuals may be interested in their own finances over time or even the progress of their own football team.

Given the following information, we can graph Sales (*y*-axis) against Time in Years (*x*-axis).

Year	Sales (in units)
1	151
2	163
3	172
4	185

The topic of interest, such as sales (given in the above example) or employment or goals scored by your favourite team is always measured from the vertical or *y*-axis.

These factors are regarded as being **dependent** upon time. Time is always measured on the horizontal or *x*-axis. In this case, we are interested in how we can *explain change over time*. A graph showing sales plotted against time is given in Figure 10.3.

Figure 10.3

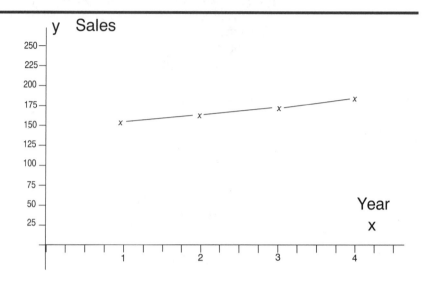

Graphs are drawn to enhance the understanding of the data. In this case we can see a steady increase over time. We should always question whether we can gain a better understanding by further analysis. Suppose we now look at the data on a quarterly basis.

Year	Quarter	Sales (in units)
1	1	20
	2	28
	3	43
	4	60
2	1	25
	2	30
	3	45
	4	63
3	1	27
	2	30
	3	48
	4	67
4	1	29
	2	32
	3	52
	4	72

In this case we can plot the changes in sales quarter by quarter as shown in the graph in Figure 10.4:

Figure 10.4

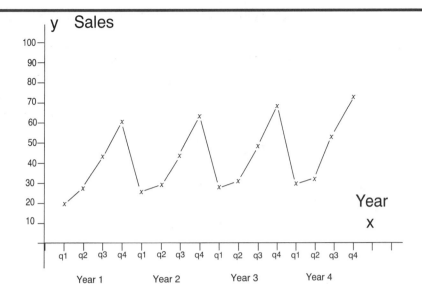

This more detailed graph allows us to observe both the general trend and the important quarterly variations. We can now clearly see that sales are predictably lower in the first quarter of each year and predictably higher in the last quarter of each year. This is an important pattern in many sets of data, such as the consumption of gas over time or the demand for Christmas goods. Clearly other goods, such as ice-cream, would have an important seasonal pattern, but rather different troughs and peaks (we typically consume more ice-cream and beer in the summer).

Exploring relationships

Plotting data against time is a special case. We can usefully plot one set of data against another set in a range of situations. Governments may want to explore the relationship between unemployment and economic activity, and companies may want to explore the relationship between sales and advertising. What is particularly important is the identification of what we want to explore or explain and what we want to use for explanation. What we are trying to explain, explore or understand is plotted on the y-axis and is referred to as the **dependent variable**. What we are using for prediction purposes is referred to as the **independent variable** or **predictor variable** and is plotted on the x-axis.

Suppose we are given the following data:

Weekly unit sales	Promotional support (£s)
80	1,500
150	1,000
200	2,200
300	2,000
450	3,000
500	3,700

In this example, a company can decide on the level of promotional support perhaps as part of a more general advertising budget. Sales will be determined in the market place and not by the company, but purchasing decisions may be influenced by the promotional material. The company will clearly be interested in this level of influence. We are therefore interested in how promotional support (the variable we control and the variable that is referred to as the independent variable) can be used to predict sales (the variable we wish to influence and that is referred to as the dependent variable). Weekly unit sales is therefore plotted against the vertical y-axis and promotional support is plotted against the horizontal x-axis as shown in Figure 10.5.

Figure 10.5

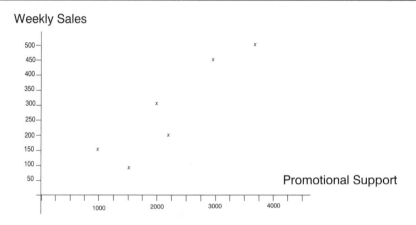

It can be seen in this case that weekly unit sales tend to increase as the level of promotional support is increased. In Chapter 11 we will consider how you can describe such relationships by using a straight line.

Exercises

Draw appropriate axes and plot (on the same graph) the points A, B, C, D and E as defined by the co-ordinates or information given in questions 1 to 5:

1 A: (25, 30)

4 D: where x is −5 and y is 10

2 B: the origin

5 E: (11,−6)

3 C: (−10,−15)

6 Draw a graph to show how the demand for cast aluminium (in tons per year) from a company of interest has changed over time using the following data:

Year	Demand for cast aluminium (in tons per year)
1995	330
1996	340
1997	320
1998	325
1999	305

7 The calls being received by a particular switchboard have been presented on a quarterly basis:

Year	Quarter 1	Quarter 2	Quarter 3	Quarter 4
1	756	673	980	1,123
2	890	677	1,123	1,343
3	899	680		

Draw a graph to show how the number of calls varies over time.

8 The following data have been collected for a particular production process on the labour time required per unit (in minutes) and the number of units produced (in thousands):

Number of units produced (in thousands)	Labour time required per unit (in minutes)
2	12
10	4
16	3
3	11
7	6
19	4

Draw a graph to show the relationship between the labour time required per unit and the number of units produced.

Worked answers

The graph required for questions 1 to 5 is shown in Figure 10.6.

Figure 10.6

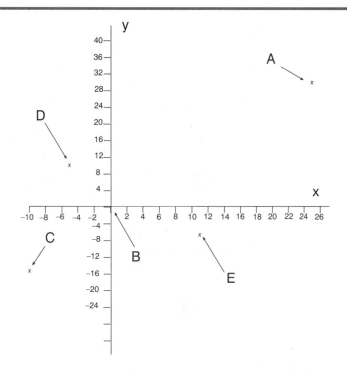

6 Demand is shown in Figure 10.7.

Figure 10.7

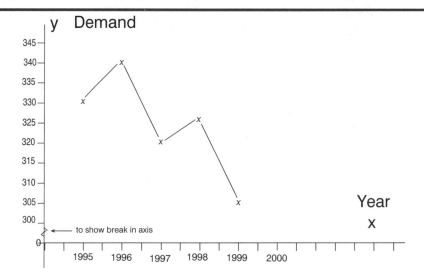

It can be seen that there has been a general but uneven fall over time. Time alone cannot explain such a trend and for a better understanding, factors such as the level of economic activity and relative price may need to be considered.

7 Number of calls is shown in Figure 10.8.

Figure 10.8

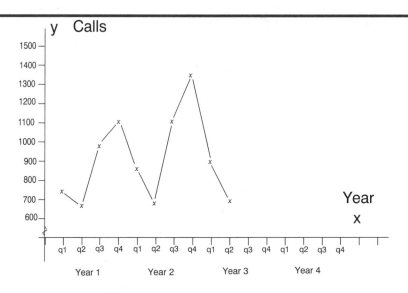

It can be seen from this graph that there has been a general increase over time, and that typically quarter 4 is higher and quarter 2 lower.

8 Labour time required is shown in Figure 10.9.

It can be seen that as the number of units produced (on the x-axis as this is the variable we can control) increases, the labour time required (on the y-axis as this is the variable that is dependent) is reduced. In this case, the scatter of points may be better described by a curve rather than a straight line.

Figure 10.9

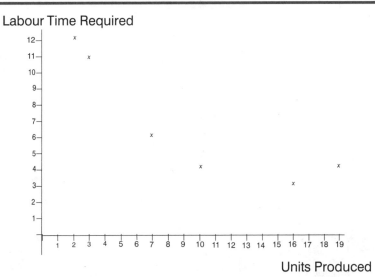

Labour Time Required

You are now ready to go on to Chapter 11, or you might wish to check your understanding of graphs by working through the re-tests in Section Six.

CHAPTER 11

Linear equations

What do the following equations all have in common?

$$y = 6$$
$$y = 4 + 3x$$
$$y = 20 - 2x$$
$$y = 5x$$

The answer is that they all **linear relationships**.

Recognizing a linear (straight line) relationship

Linear relationships will appear as straight lines on a graph. In general, a linear relationship will take the form:

$$y = a + bx$$

where x values are plotted against the horizontal axis,
     y values (corresponding to the x values) are plotted on the vertical axis,
     a is a constant term, and
     b is referred to as the gradient.

Considering each of these equations in turn:

$\underline{y = 6}$ In this case $a = 6$ and $b = 0$.

As you can see in the graph (Figure 11.1), a horizontal line cutting the vertical axis at 6 is produced.
     No matter what the value of x, the value of y will always equal 6.

$\underline{y = 4 + 3x}$ In this case $a = 4$ and $b = 3$.

To plot this linear relationship we only need two points (but we can use more to prove a point!) which can then be joined by a ruler line. You can use any two points but generally those at the extreme ends of the range are used to improve drawing accuracy. Using the range from $x = -5$ to $x = 5$, then we may as well use these values:

     for $x = -5$, $y = 4 + 3 \times (-5) = 4 - 15 = -11$ (point A)

     for $x = 5$, $y = 4 + 3 \times 5 = 4 + 15 = 19$ (point B)

Figure 11.1

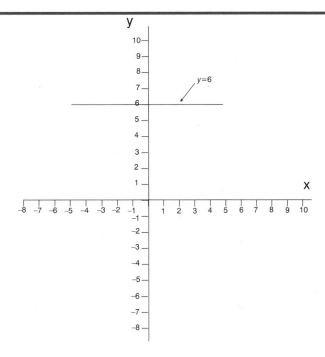

Figure 11.2

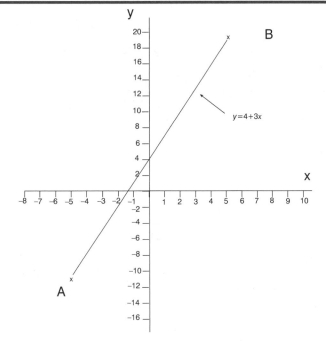

Points A and B joined by a 'ruler' line are shown on the graph in Figure 11.2.
The change in the value of y corresponding to the changes in the value of x are often
shown by means of a table as below:

x	$3x$	$y = 4 + 3x$
−5	−15	−11
−4	−12	−8
−3	−9	−5
−2	−6	−2
−1	−3	1
0	0	4
1	3	7
2	6	10
3	9	13
4	12	16
5	15	19

There are two important points to note:

1. The line on the graph will cut the vertical y-axis at the point given by a in the general equation. We have already referred to this as the **constant term** (in this case it is 4),

2. For each unit increase in x (as x goes from −5 to −4 to −3 and so on) there is a corresponding constant change in y. This **gradient** is defined by b in the general equation. In this example, for each unit increase in x there is a 3 unit increase in y (y goes from −11 to −8 to −5 through to 13, 16 and 19).

$$y = 20 - 2x \qquad \text{In this case } a = 20 \text{ and } b = -2.$$

Again we only need two points to plot a straight line. Using the range from $x = -5$ to $x = 5$, we can use the extreme values:

for $x = -5$, $y = 20 - 2 \times (-5) = 20 + 10 = 30$ (point A)

for $x = 5$, $y = 20 - 2 \times 5 = 20 - 10 = 10$ (point B)

We can join points A and B to show a straight line graph (Figure 11.3):

Figure 11.3

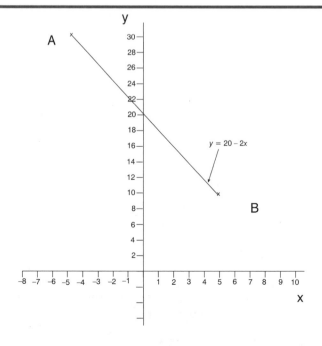

Here the constant term is 20 and the gradient −2. When a line is going down from the left to the right we refer to a **negative gradient**. For each unit increase in x the value of y will drop by 2 units.

Another example:

$$y = 5x \qquad a = 0 \text{ and } b = 5.$$

For this example we use the range from $x = -10$ to $x = 10$ to plot the graph. Using two convenient values, we have:

for $x = -10$, $y = 5(-10) = -50$ (point A)

for $x = 10$, $y = 5 \times 10 = 50$ (point B)

The graph is shown in Figure 11.4.

Figure 11.4

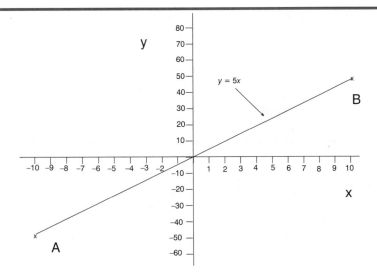

As you can see, the line goes through the origin (0,0) and we have a **positive gradient** of 5.

Finding the linear equation given a straight line graph

So far, we have looked at plotting an equation on a graph. Now reverse this, and consider how we can determine the equation given a set of points on a graph. We can plot the points given below in the usual way:

x	y
−2	−7
0	−3
1	−1
4	5

The graph corresponding to the given data is shown in Figure 11.5.

Figure 11.5

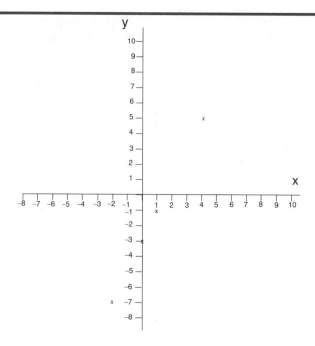

In this case we can easily draw a line through the points (Figure 11.6).

Figure 11.6

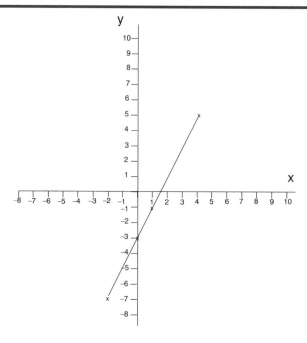

We now need to determine the values for a and b. If we return to our general form of a linear relationship $y = a + bx$ we can see that two values, or parameters, need to be determined.

The determination of a
We know that when $x = 0$, $y = a$. So the value of a, called the intercept, can be found at the point where the straight line cuts the y-axis (see Figure 11.7).

The determination of b
The value of b gives the increase in y corresponding to a unit increase in x (a decrease would be given as a negative value). To find this graphically, we draw an appropriate right-angled triangle and divide the height by the base (see Figure 11.7).

Figure 11.7

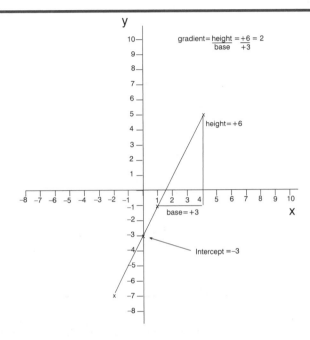

It should be noted that a larger of smaller triangle could have been used and the gradient, which is concerned only with the change in y resulting from a unit change in x, would stay the same. In this case, the equation that describes the line joining the points is

$$y = -3 + 2x.$$

As a second example, consider the following set of data:

x	y
−5	29
0	4
5	−21
10	−46

The graph (Figure 11.8) shows a plot of the points given, a fitted straight line, the intercept (value of a) and the determination of the gradient (the value b).

Figure 11.8

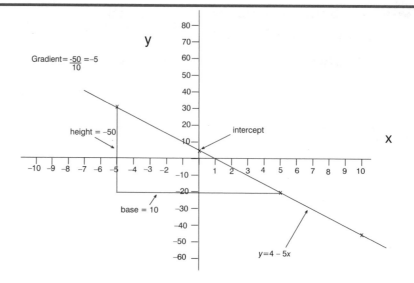

The equation in this case is

$$y = 4 - 5x.$$

Exercises

Draw appropriate axes and on the same graph plot the following:

1 $y = 37 + 0.2x$ between $x = -10$ and $x = 15$

2 $y = 10 + 2x$ between $x = -20$ and $x = 15$

3 $y = -26 + 0.2x$ between $x = -20$ and $x = 5$

4 Using appropriate axes, plot the following points:

x	y
−10	37
−5	17
−4	13
−1	1
0	−3

5 On the graph you have just drawn (question 4), join the points with a straight line. Now determine the linear equation that describes the line.

Worked answers

The graph required for questions 1 to 3 is shown in Figure 11.9.

Figure 11.9

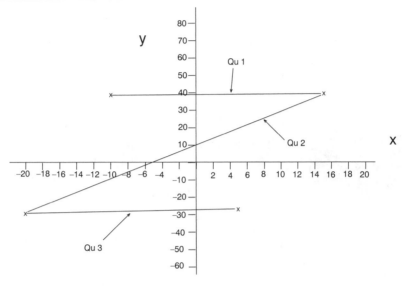

Did you get the Z shape? It is worth checking carefully the corner points for correctness. Did you notice that the first and third equations produced parallel lines (they both had a gradient of 0.2).

The graph (Figure 11.10) shows the plotting of points from questions 4 and the determination of the equation $y = -3 - 4x$ for question 5.

Figure 11.10

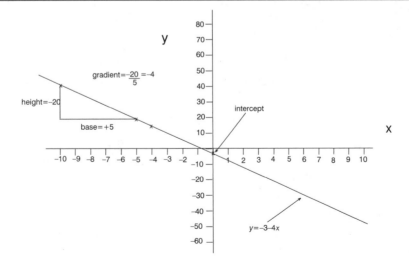

You are now ready to go on to Chapter 12, or you might wish to check your understanding of linear equations by working through the Re-tests in Section Six.

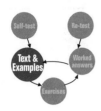

CHAPTER 12

Simultaneous equations

A pair of simultaneous equations represent two conditions and will be true at the same time (i.e. simultaneously), if and when they cross each other.

Solving simultaneous equations graphically

As we have seen, we are able to plot one, two or more equations on a single graph. The point at which the lines cross is of particular interest to us. This point gives values that satisfy both equations at the same time or simultaneously. Suppose we are working with the following equations:

$$y = 20 - 2x$$

and

$$y = 2 + 2x$$

They are both linear (y is a function of x but not x^2 or x^3). We only need to determine two points for each equation to show them as straight line graphs. Choosing $x = 0$ and $x = 10$ gives:

$y = 20 - 2x$ when $x = 0$, $y = 20$
and
when $x = 10$, $y = 0$

$y = 2 + 2x$ when $x = 0$, $y = 2$
and
when $x = 10$, $y = 22$

It can be seen from Figure 12.1 that the lines cross when $x = 4.5$ and $y = 11$.

When $x = 4.5$ both equations will have a y value of 11. (Therefore both equations are true at this point.)

Figure 12.1

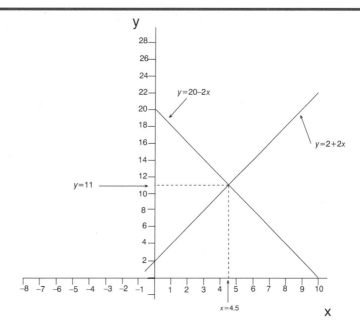

Solving simultaneous equations by elimination

We are sure you will agree that it would soon become tedious to solve all problems of this kind using the graphical method and it would often lack the required accuracy. Solving simultaneous equations by elimination is a method you may recall from school. Essentially, **the equations are scaled in such a way that the coefficients of x or y are the same and then add or subtract the equations to eliminate x or y.** An example will make it clearer.

Given the equations $y = 20 - 2x$ and $y = 2 + 2x$, we can easily eliminate x by adding them:

$$\begin{aligned} y &= 20 - 2x \\ y &= 2 + 2x \end{aligned} \quad + $$
$$\overline{2y = 22}$$

If $2y = 22$, then $y = 11$.

Once we know that $y = 11$, we can substitute this value into either equation to find the value of x. Using the first equation for convenience:

$$11 = 20 - 2x$$
$$-9 = -2x$$
$$x = 4.5$$

Often simultaneous equations are presented in a slightly different form with the x and y terms both on one side and the number term on the other. Suppose we are required to solve the following:

$$\text{and} \quad \begin{aligned} -3x + 4y &= 5 \\ 2x - 5y &= -15 \end{aligned}$$

We can either multiply the equations in such a way as to eliminate x or to eliminate y.

Eliminate x

Multiply the first equation by 2 and the second equation by 3. As you can see, when we do this the coefficient of x in the first equation is -6 and in the second equation is $+6$.

$$-6x + 8y = 10$$
$$6x - 15y = -45$$

Now add the two equations together:

$$-6x + 8y = 10$$
$$6x - 15y = -45 \quad +$$
$$\overline{-7y = -35}$$
$$y = 5$$

We can then substitute $y = 5$ in to either the first or the second equation. Using the first we get

$$-3x + 4 \times 5 = 5$$
$$-3x + 20 = 5$$
$$-3x = -15$$
$$x = 5$$

In this case x and y both equal 5.

Eliminate y

Alternatively, we can eliminate y: multiply the first equation by 5 and the second equation by 4 so that y would have the coefficient $+20$ in the first equation and -20 in the second equation. Then adding the two equations together allows us to find that $x = 5$. By substitution, we also discover that $y = 5$.
It does not matter which you choose.

Solving simultaneous equations by substitution

This essentially requires two steps.

First we use one equation to put y in terms of everything else.

Secondly we substitute the value we have for y into the other equation and then rearrange to tidy.

It is easier to do than explain! (N.B. It is just as valid to find x in terms of everything else, and is sometimes easier.)
Given the equations:

$$y = 20 - 2x$$
$$\text{and} \quad y = 2 + 2x$$

we already have y in terms of x (in both equations) and can substitute directly from the first equation into the second equation:

$$(20 - 2x) = 2 + 2x$$
$$18 = 4x$$
$$x = 4.5$$

Since $x = 4.5$, we can substitute this value into the first equation (or we could use the second equation) to find

$$y = 20 - 2 \times 4.5$$
$$= 20 - 9$$
$$= 11$$

Just for illustration, suppose we decided to find x in terms of y in the first equation and then substitute into the second equation (we know the answer will be the same but it does usefully show some of the algebra that can be involved).

Still working with the equations $y = 20 - 2x$ and $y = 2 + 2x$, we can use the first equation to put x in terms of y:

$$-2x = y - 20$$

Dividing by -2 gives

$$x = -\frac{y}{2} + 10$$

Substituting this value for x into the second equation gives

$$y = 2 + 2\left(-\frac{y}{2} + 10\right)$$
$$y = 2 - y + 20$$
$$2y = 22$$
$$y = 11$$

Taking this value of y back into the first equation gives

$$11 = 20 - 2x$$
$$-9 = -2x$$
$$x = 4.5$$

We did mention in the previous section that simultaneous equations are often presented in a slightly different form with the x and y terms both on one side and the number term on the other. Let us again solve the following equations but this time by the substitution method.

Given

$$-3x + 4y = 5$$
$$\text{and} \quad 2x - 5y = -15$$

we must first decide whether to put x in terms of y or y in terms of x. It does not really matter but generally we will look for the approach that **makes the arithmetic easier**. Given that there are no obvious advantages, we will put y in terms of x in the first equation and then substitute into the second equation.

So

$$-3x + 4y = 5$$
$$4y = 5 + 3x$$
$$y = \frac{5}{4} + \frac{3}{4}x$$

Substituting this into the second equation

$$2x - 5\left(\frac{5}{4} + \frac{3}{4}x\right) = -15$$
$$2x - \frac{25}{4} - \frac{15}{4}x = -15$$

At this stage it is probably easier to multiply through by 4

$$8x - 25 - 15x = -60$$
$$-7x = -35$$
$$x = 5$$

Taking this x value back to the first equation (but it could just as easily be the second)

$$-3 \times 5 + 4y = 5$$
$$-15 + 4y = 5$$
$$4y = 20$$
$$y = 5$$

Problem solving with simultaneous equations

Simultaneous equations are not just included here to test your algebra. They do have practical applications. Consider the following example.

A company operates two types of aircraft, the RS101 and the JC111. The RS101 is capable of carrying 40 passengers and 30 tons of cargo, whereas the JC111 is capable of carrying 60 passengers and 15 tons of cargo. You need to find an appropriate combination of the two types of aircraft to carry 480 passengers and 180 tons of cargo.

If x is used to denote the number of RS101s and y is used to denote the number of JC111s then we can express the problem in the form of two equations.

The passenger equation: $40x + 60y = 480$

The cargo equation: $30x + 15y = 180$

We can see for example, that if the RS101 can take 40 passengers then 40 times the number of RS101s or $40x$ will give the total number of passengers taken by that type of aircraft. To solve this problem as specified, we can multiply the second equation by 4 and take it away from the first equation.

$$
\begin{array}{rl}
40x + 60y & = 480 \; - \\
\underline{120x + 60y} & \underline{= 720} \\
-80x \quad\quad & = -240 \\
x \quad\quad & = 3
\end{array}
$$

We can take this solution back into the first equation (or the second if we choose):

$$
\begin{array}{rl}
40 \times 3 + 60y & = 480 \\
120 \quad + 60y & = 480 \\
60y & = 360 \\
y & = 6
\end{array}
$$

A suitable combination of aircraft would be 3 RS101 and 6 JC111.

To make the problem more realistic we would also need to consider the potential profits or costs associated with the different types of aircraft. We would also need to relax the restriction that all of the passenger and cargo capacity was fully utilized. In addition we would also need to consider other factors, such as the availability of aircraft. This example is more fully considered in *Quantitative Methods for Business Decisions* by Curwin and Slater.

Exercises

1 Given the following equations
$$y = 3x - 2$$
and $y = -2x + 13$

(a) plot both on the same graph and identify the point of intersection

(b) solve as simultaneous equations.

Solve each of the following pairs of simultaneous equations using appropriate algebra:

2 $2x + y = 10$
$3x - 2y = 15$

3 $4y - 2x = 44$
$2y + 3x = 22$

4 $-3x + 4y = -1$
$5x + 7y = -12$

5 $2x + 3y = 20$
$-x + 2y = 18$

6 Suppose you have been given the following relationships to describe the supply of goods to a market in terms of price (p) and quantity (q), and the demand for goods in the same market also in terms of p and q.

Following the tradition of economics, price is expressed as a function of quantity:

demand: $p = 350 - \frac{1}{3}q$
and
supply: $p = 100 + \frac{1}{2}q$

The point at which the price and quantity is equal is known as the equilibrium point.

(a) show the demand and supply functions graphically

(b) solve as simultaneous equations to find the equilibrium price and quantity.

Figure 12.2

Worked answers

1 (a)

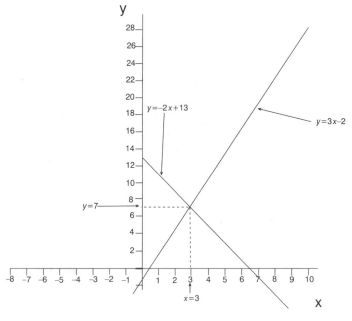

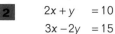

(b) $y = 3x - 2$

$y = -2x + 13$ *just eliminate y*

$3x - 2 = -2x + 13$

$5x = 15$

$x = 3$ *Substitute into either of the equations to find y. Using the first equation*

$y = 3 \times 3 - 2$

$= 7$

The answer $x = 3$ and $y = 7$.

2 $2x + y = 10$

$3x - 2y = 15$ *multiply the first equation by 2 and add*

$4x + 2y = 20$

$\underline{3x - 2y = 15}$ $+$

$7x = 35$ *By substituting $x = 5$ into the first equation (or we could use the second)*

$x = 5$

$$2 \times 5 + y = 10$$

$$y = 10 - 10$$

$$y = 0$$

The answer is $x = 5$ and $y = 0$

3 $\quad 4y - 2x = 44$

$2y + 3x = 22$ multiply the second equation by 2 and subtract

$$4y - 2x = 44$$
$$\underline{4y + 6x = 44} \quad ^-$$
$$-8x = 0$$
$$x = 0$$

substituting $x = 0$ into the first equation

$$4y - 2 \times 0 = 44$$

$$4y = 44$$

$$y = 11$$

The answer is $x = 0$ and $y = 11$

4 $\quad -3x + 4y = -1$

$5x + 7y = -12$

multiply the first equation by 5 and the second equation by 3, and then add

$$-15x + 20y = -5$$
$$\underline{15x + 21y = -36} \quad ^+$$
$$41y = -41$$
$$y = -1$$

substituting $y = -1$ in to the first equation

$$-3x + 4 \times (-1) = -1$$

$$-3x - 4 = -1$$

$$-3x = -1 + 4$$

$$-3x = 3$$

$$x = -1$$

The answer is $x = -1$ and $y = -1$.

5 $\quad 2x + 3y = 20$ multiply the second equation by 2 and add

$-x + 2y = 18$

$$2x + 3y = 20$$
$$\underline{-2x + 4y = 36} \quad ^+$$
$$7y = 56$$
$$y = 8$$

substituting $y = 8$ in to the first equation

$$2x + 3 \times 8 = 20$$

$$2x + 24 = 20$$

$$2x = -4$$

$$x = -2$$

The answer is $x = -2$ and $y = 8$.

6 (a)

Figure 12.3

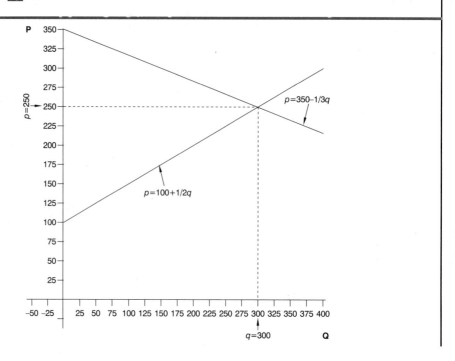

(b) Given the two equations for supply and demand:

demand: $p = 350 - 1/3\,q$

supply: $p = 100 + 1/2q$

$350 - \frac{1}{3}q = 100 + \frac{1}{2}q$ *eliminating p*

$2100 - 2q = 600 + 3q$ *multiplying through by 6*

$-5q = -1500$

$q = 300$

substituting q = 300 into the first equation

$p = 350 - \frac{1}{3} \times 300$

$= 350 - 100$

$= 250$

The answer is $q = 300$ and $p = 250$.

Mathematics like this are used in economics to consider market equilibrium.

You are now ready to go on to Chapter 13, or you might wish to check your understanding of simultaneous equations by working through the Re–tests in Section Six.

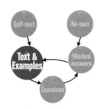

CHAPTER 13

Quadratic equations

So far we have only considered linear equations; those equations that produce straight line graphs. Equations of a particular kind we often refer to as a family. Linear equations are one of many families. Equations can be specified not only in terms of x but also x^2 or x^3 or x^4 or (x and z) or (x and x^2 and z and z^2) or any infinite variety of ways.

Recognizing quadratic equations

In this chapter, we explore one of the many families that exist: the quadratic equation.

A quadratic equation is defined by the form:

$$y = ax^2 + bx + c$$

where x values are plotted on the horizontal x-axis,

y values (corresponding to the x values) are plotted on the vertical y axis,

a, b, c are constants

and $a \neq 0$ (the x^2 term must exist within the equation).

The following are both examples of quadratic equations:

$$y = x^2 - 8x + 12$$

and

$$y = -x^2 + 7x - 10$$

Given that the first equation will produce a $\cup$ shape and the second will produce an $\cap$ shape, we can no longer just use two points and join with a ruler. We now need to produce a curve using sufficient points.

Many people prefer to go directly from each x value to the corresponding y value (which is relatively easy using a calculator that will carry forward the mathematical logic or a spreadsheet) but others prefer to build-up their answer using a table format. We will demonstrate the use of the table format here and leave you to judge whether this is useful or necessary for you.

Plotting
$y = x^2 - 8x + 12$

A table is constructed to work through each of the parts of the equation which can then be combined. Taking values from 0 to 10, we can work out the values for x^2 first.

x	0	1	2	3	4	5	6	7	8	9	10
x^2	0	1	4	9	16	25	36	49	64	81	100

In a similar way, we can work out $-8x$ for the same range of x values:

x	0	1	2	3	4	5	6	7	8	9	10
$-8x$	0	−8	−16	−24	−32	−40	−48	−56	−64	−72	−80

and, of course, we know that the last bit of the function is always equal to $+12$. We can now put all these parts together to find y for the given values of x.

x	0	1	2	3	4	5	6	7	8	9	10
x^2	0	1	4	9	16	25	36	49	64	81	100
$-8x$	0	−8	−16	−24	−32	−40	−48	−56	−64	−72	−80
$+12$	+12	+12	+12	+12	+12	+12	+12	+12	+12	+12	+12
y	+12	+4	0	−3	−4	−3	0	+5	+12	+21	+32

We can now plot the points of y against x. We need to join these points with a smooth curve rather than a series of straight lines – see Figure 13.1:

Figure 13.1

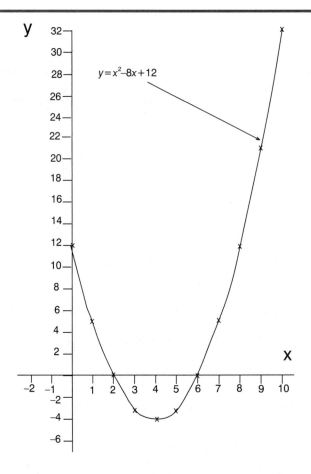

$y = x^2 - 8x + 12$

This graph will produce the $\cup$ shape because of the dominance of $+x^2$. As positive and negative x values get larger, $+x^2$ will produce proportionately larger and larger positive values (remember that negatives squared will give a positive).

Plotting
$y = -x^2 + 7x - 10$

We can also build–up a table for this equation:

x	0	1	2	3	4	5	6	7	8	9	10
$-x^2$	0	−1	−4	−9	−16	−25	−36	−49	−64	−81	−100
$+7x$	0	7	14	21	28	35	42	49	56	63	70
-10	−10	−10	−10	−10	−10	−10	−10	−10	−10	−10	−10
y	−10	−4	0	2	2	0	−4	−10	−18	−28	−40

We can now plot the values of y against x and join with a smooth curve (Figure 13.2).

Figure 13.2

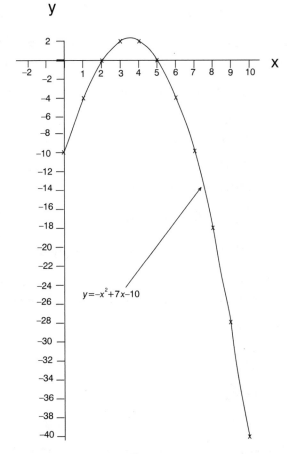

The graph will produce a $\cap$ shape because of the dominance of $-x^2$. As positive and negative x values get larger, $-x^2$ will produce proportionately larger and larger negative values.

Roots

The roots of an equation are the points where the graph of the equation crosses the x-axis, i.e. where $y = 0$. It can be seen from the previous graphs that

the roots for $\quad y = x^2 - 8x + 12 \quad$ are $x = 2$ and $x = 6$ from Figure 13.1

and

the roots for $\quad y = -x^2 + 7x - 10 \quad$ are $x = 2$ and $x = 5$ from Figure 13.2.

Roots can also be found by factorization or by use of a formula – both of which we will now consider.

Factorizing

This is an approach that many prefer and that many can still remember from school. It has the quality and challenge of a crossword puzzle **but** it should be remembered that not all quadratic equations can be easily factorized.

To begin with, let's expand out the expression

$$(x + a)(x + b)$$

We need to multiply the contents of the second bracket by x **and** by $+a$:

second bracket multiplied by x: $\qquad\qquad x^2 + bx$

second bracket multiplied by a: $\qquad\qquad ax + ab$

if we sum these to get

second bracket multiplied by $(x + a)$: $\qquad x^2 + bx + ax + ab$

Which can be more neatly written as: $\qquad\qquad x^2 + (a + b)x + ab$

Turning this around, tells us that a quadratic can be factorized into a pair of brackets:

$$x^2 + (a + b)x + ab = (x + a)(x + b)$$

A couple of examples will clarify the usefulness of this result.

Factorizing $y = x^2 - 8x + 12$

We first need to set $y = 0$. The quadratic becomes $x^2 - 8x + 12 = 0$.
  To find the values of a and b that will make $x^2 - 8x + 12 = (x + a)(x + b)$, we need

$$a + b = -8$$

and

$$ab = 12$$

(A consequence of the result above)

To restate, we need a couple of numbers that when added together will equal -8 and when multiplied together will equal 12. Would you agree that these numbers must be -2 and -6?

We get $\qquad\qquad x^2 - 8x + 12 = (x - 2)(x - 6) = 0$

If $\qquad\qquad (x - 2)(x - 6) = 0$

then $(x - 2) = 0$
                              $(x - 6) = 0$

The roots of the quadratic are $x = 2$ and $x = 6$ (as shown on the earlier graph in Figure 13.1).

Factorizing
$y = -x^2 + 7x - 10$

We first need to set $y = 0$. The quadratic becomes $-x^2 + 7x - 10 = 0$.
   In this case, it is probably easier to multiply through by -1 first to get:

$$x^2 - 7x + 10 = 0$$

We now need a couple of numbers that when added together will equal -7 and when multiplied together will equal 10. Would you agree that these numbers must be -2 and -5?

We get $x^2 - 7x + 10 = (x - 2)(x - 5) = 0$

If $(x - 2)(x - 5) = 0$

then $(x - 2) = 0$
                              $(x - 5) = 0$

The roots of the quadratic are $x = 2$ and $x = 5$ (as shown in Figure 13.2).

Solution by
formula

The alternative to using the factorizing method is to use a formula. This again may be something you remember from school. Many prefer this method because it always works and it gives the answer without having to puzzle out the use of double brackets. The formula works with the standard quadratic equation:

$$ax^2 + bx + c = 0$$

and the roots are found by using

$$x = \frac{-b \pm \sqrt{b^2 - 4ac}}{2a}$$

Finding the roots
of $x^2 - 8x + 12 = 0$

$$a = 1 \quad b = -8 \quad c = 12$$

(You need to remember that x^2 means
$+ 1$ times x^2)

Substituting these values into the equation gives:

$$x = \frac{-(-8) \pm \sqrt{(-8)^2 - 4 \times 1 \times 12}}{2 \times 1}$$

$$x = \frac{8 \pm \sqrt{(64 - 48)}}{2}$$

$$x = \frac{8 \pm \sqrt{16}}{2}$$

$$x = \frac{8 \pm 4}{2}$$

$$x = \frac{12}{2} \text{ or } \frac{4}{2}$$

The roots of the quadratic are $x = 6$ or $x = 2$

Finding the roots of $-x^2 + 7x - 10 = 0$

So

$$a = -1 \qquad b = +7 \qquad c = -10$$

Substituting these values into the equation gives:

$$x = \frac{-7 \pm \sqrt{\left((7)^2 - 4 \times (-1) \times (-10)\right)}}{2 \times (-1)}$$

$$x = \frac{-7 \pm \sqrt{(49 - 40)}}{-2}$$

$$x = \frac{7 \pm \sqrt{9}}{-2}$$

$$x = \frac{-7 \pm 3}{-2}$$

$$x = \frac{-10}{-2} \text{ or } \frac{-4}{-2}$$

The roots of the quadratic are $x = 5$ or $x = 2$

Exercises

1 Draw the graph of $y = x^2 + 5x + 4$ from $x = -5$ to $x = 10$ and identify the roots.

Using the method of factorization, find the roots of the following quadratic equations:

2 $x^2 - 9x + 18 = 0$

3 $x^2 + 8x + 12 = 0$

4 $-x^2 + 49 = 0$

5 $-x^2 + 6x + 27 = 0$

Using the equation method solve the following quadratic equations

6 $x^2 - 3.2x + 2.55 = 0$

7 $x^2 - 0.8x - 3.2625 = 0$

Worked answers

1

Figure 13.3

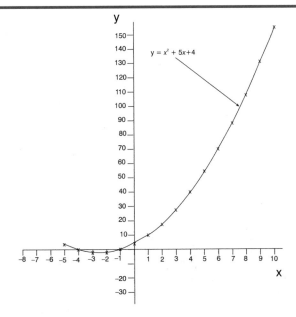

$y = x^2 + 5x + 4$

2 $x^2 - 9x + 18 = 0$ $a+b=9, ab=18$
$(x - 3)(x - 6) = 0$

The roots are $x = 3$ and $x = 6$

3 $x^2 + 8x + 12 = 0$
$(x + 2)(x + 6) = 0$

The roots are $x = -2$ and $x = -6$

4 $-x^2 + 49 = 0$ *watch for the $-x^2$*
$(-x + 7)(x + 7) = 0$

The roots are $x = -7$ and $x = 7$

5 $-x^2 + 6x + 27 = 0$ *you could also multiply through by −1 first*
$(-x - 3)(x - 9) = 0$ *and then factorize*

The roots are $x = -3$ and $x = 9$

6 $x^2 - 3.2x + 2.55 = 0$

$a = 1$ $b = -3.2$ $c = 2.55$

$$x = \frac{-(-3.2) \pm \sqrt{\left((-3.2)^2 - 4 \times 1 \times 2.55\right)}}{2 \times 1}$$ *substituting*

$$x = \frac{3.2 \pm \sqrt{(10.24 - 10.2)}}{2}$$

$$x = \frac{3.2 \pm \sqrt{0.04}}{2}$$

$$x = \frac{3.2 \pm 0.2}{2}$$

The roots of the quadratic are $x = 1.7$ or $x = 1.5$

7 $x^2 - 0.8x - 3.2625 = 0$

$a = 1$ $\qquad b = -0.8$ $\qquad c = -3.2625$

Substituting these values into the equation gives:

$$x = \frac{-(-0.8) \pm \sqrt{\left((-0.8)^2 - 4 \times 1 \times (-3.2625)\right)}}{2 \times 1}$$ *substituting*

$$x = \frac{0.8 \pm \sqrt{(0.64 + 13.05)}}{2}$$

$$x = \frac{0.8 \pm \sqrt{13.69}}{2}$$

$$x = \frac{0.8 \pm 3.7}{2}$$

The roots of the quadratic are $x = -1.45$ or $x = 2.25$

You are now ready to go on to the spreadsheet section, but, if you have not already done so, it would be a good idea to work through Section Six to check your understanding of equations and graphs.

Section 5

Using Spreadsheets

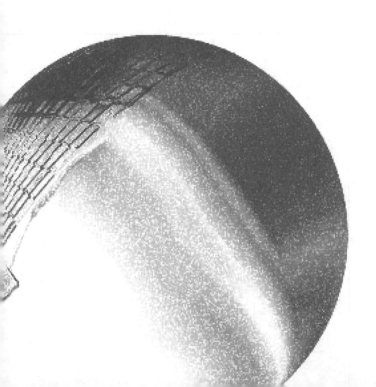

CHAPTER 14

Spreadsheets

Spreadsheets have completely revolutionized the way we do sums. Even more, they are used daily by people who consider themselves *not particularly good at maths* to calculate the results of complex relationships between variables. The bottom line is that spreadsheets help them with what they do; in some cases, even making it easy. To get the most out of a spreadsheet, you need to understand some of the basics we have dealt with earlier in this book – things like the order of operations. Also, having at least a rough idea of the likely magnitude of the answer helps avoid plainly embarrassing mistakes.

There are at least three different ways to use a spreadsheet. First, you could just use the built-in functions, like adding up a set of figures, or working out a regression relationship. Secondly, you could use spreadsheets and templates prepared by somebody else, such as the accounts department or a commercially produced add-in. Both of these are useful and valuable in the right context, but they limit you to only being able to deal with particular situations. To be able to deal with any situation which arises, you need to use spreadsheets in a third way, that is, to build your own spreadsheets from scratch so that you can incorporate all of the relevant factors. We are not suggesting that you continually re-invent the wheel – far from it. Where there are pre-written routines which meet our needs, we should use them; but where there are none, we should be able to write our own.

We are assuming that you have access to a spreadsheet and that you can start it by clicking on the appropriate icon on the screen. For simplicity we are going to be working with Excel97 out of the Microsoft Office97 package, but most of the things you find here will apply to any spreadsheet. Once you have the spreadsheet open on the computer screen it is a good idea to maximize it (so that it fills the screen) since this makes it easier to work with your data.

A blank spreadsheet

You can only see a small part of the spreadsheet on the screen at any one time. While looking at a blank spreadsheet it is worth examining its structure. Notice that the spreadsheet is divided into both columns and rows and that the columns are labelled with letters (A, B, C, etc.) and that the rows are labelled with numbers (1, 2, 3, etc.) This means that any individual **cell** on the spreadsheet can be identified **uniquely** by specifying the *Column* and the *Row* (see Figure 14.1). You might like to find out how many columns and rows there are – **a lot!** – but you can always add more should you need them. With Excel there are in fact three spreadsheets within the workbook you have opened – look at the bottom left of the screen and you will see three tabs. The reason for this is that when you build particularly complicated spreadsheets, it is often useful to split up sections of the calculations on to different sheets, and then link them up at the end.

Figure 14.1
A blank spreadsheet

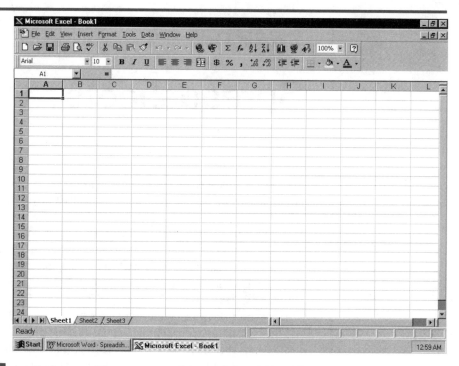

Cells

A cell of a spreadsheet can contain one of three things. These are:

words

numbers

formulae

It is important to put **words** into your spreadsheet to help you remember what you have done, or are trying to do. It is also useful if you show the spreadsheet to someone else. To put a word into a cell, just click on the cell you want and type in the word or words. Do not worry if you appear to go over into an adjoining cell at this point.

Doing things with **numbers** is the reason you started using a spreadsheet in the first place. Again, to put a number into a cell, just click on the cell you want and type in the number.

Formulae are the clever bit! Rather than simply multiplying say 6 times 4, we put the six in one cell, the 4 in another cell, and then tell the spreadsheet to multiply the two **cells** together. For example, if we put 6 into cell A4, and 4 into cell A5, then in cell A7 (say) we could put

=A4*A5 – which means multiply them together

This does not look like an improvement at first, but if we decide that we now want to multiply 6 by 27.8, all we have to do is change cell A5 from 4 to 27.8, and the answer will **automatically appear** in cell A7. It is this feature, in particular, which makes spreadsheets so useful.

Building a spreadsheet

To illustrate the various aspects of constructing a spreadsheet, we will use the following scenario. Suppose that you have been given a sum of money by a loving aunt, who says:

> You need a car at university, well at least you did in my day! I'll give you £4,500 to get you going.

You now need to decide on how much you can afford to spend on the car itself, what it will cost to run, whether you can keep it going for three years and the possible repair costs. (We appreciate that if you found yourself in this very fortunate position, you may just react by buying something, but the construction of the spreadsheet will illustrate many points.) For best results, we suggest that you actually build this spreadsheet along with reading the explanation in the text.

Putting a title at the top

Start your spreadsheet package and go to cell A1 in the top left-hand corner of the screen. Type in the following text:

> Buying and Running a Car

And then press the RETURN key. You will see the text appear on the screen, and also find that it goes over into cell B1 and C1. (This doesn't really matter for a title.) All of the text is stored in cell A1, so if we want to change it, we must go back to that cell, and **not** to B1 or C1. Go back to A1 now, and make the text bold, by clicking on the B icon. Change the size of the text to 18 from whatever it is (probably 10 or 12). This makes the title stand out, and also makes it overlap into cell E1.

Your spreadsheet should now look something like Figure 14.2.

Figure 14.2

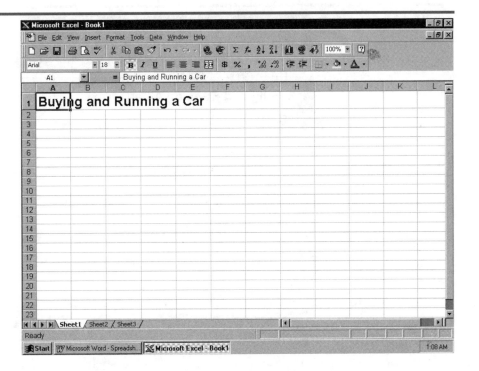

Some more words

In this scenario we can distinguish three elements to the costs of the project – the one-off cost of purchasing the car, the (fixed) annual costs, and the variable running costs. The latter group will be related to the annual milage covered and the economy of the vehicle.

On your spreadsheet go to cell A3 and put in the words 'One-Off Cost'; then to cell A7 and put in 'Annual Costs' and then to cell A13 and put in 'Other Running Costs'. Make each of these cells bold and 12 point type.

Then add the following:

Cell	Words		Cell	Words
A4	Price		C8	Next Year
A9	Car Tax		D8	Year After
A10	Insurance		B14	This Year
A11	MOT		C14	Next Year
A15	Petrol		D14	Year After
A16	Oil		H2	This Year
A17	Exhaust		I2	Next Year
A18	Servicing		J2	Year After
A19	Tyres		G3	Annual Milage
A22	Total Cost		G4	Petrol Cost
B8	This Year		G7	Miles per litre

Don't worry if some of the cells seem to overlap. Your spreadsheet should now look like Figure 14.3.

Figure 14.3

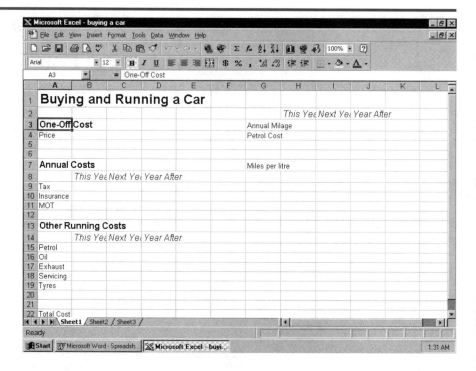

Don't forget to save your spreadsheet regularly, say every 10 minutes.

Numbers

We will begin by putting in some initial data. For simplicity we will again just specify which numbers to put into each cell as a table.

Cell	Number		Cell	Number
B4	2000		H4	0.8
B9	150		H7	8
B10	350		B18	100
B11	33		C19	60
H3	5000		D17	40

These data form the basis for the subsequent calculations.

Save your spreadsheet again now.

Formulae

It is the use of formulae which make spreadsheets so useful and so adaptable. In our case we will just use a few basic ideas.

Go to cell G9 and put in the words 'Total Petrol' and then go to cell H9. In this cell we want to find the number of litres of petrol purchased this year. This will be equal to the total mileage divided by the number of miles per litre. The total milage this year is in cell H3, and the Miles per Litre is in cell H7, so we need to put the formula

$$=H3/H7$$

into cell H9. Note the = sign which indicates that this is a formula. You should see the result 625 appear. We know the cost of petrol is in cell H4. In cell B15 we want the total cost of petrol for the year, so if we multiply the cost per litre by the amount, we will get the result. Put the formula

$$=H4*H9$$

into cell B15. You should get the result of 500. (Note that an asterisk (*) is used for multiplication.)

Now go to cell B12 and we will use the summation function. This is something which is built into all spreadsheets and allows you to add up a column or row of data. In Excel97 make sure you have B12 selected and then look for a Σ (sigma) sign on the toolbars at the top of the screen. Point to this with the mouse and click once. You will see the formula

$$=SUM(B9:B11)$$

appear and the cells B9, B10 and B11 show as a highlighted region on the spreadsheet. These are exactly the cells we want to add together, so we can just click once on the green arrow next to the formula (near the top of the screen, just below the toolbars). The result 533 appears in cell B12. If we had wanted to add more or less cells together we could easily modify the formula; as you can see, in the brackets you need to specify the first and last cells that you want to add, and the spreadsheet will then add those and any in between.

Go to cell B20 and again click on sigma. This time the computer does not select the cells we want, so you will need to change the formula to

=SUM(B15:B19)

and then you will get the result 600.

To get the costs for the year, go to cell B22 and put in a formula to add the One-Off Cost, the Annual Costs and the Other Running Costs. This will be

=B4+B12+B20

The result is 3133.

Figure 14.4

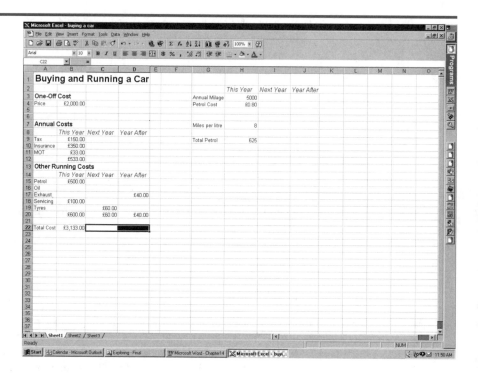

To make sure your spreadsheet works, try changing cell H3 to 10000 and make sure the result in cell B22 comes to 3633. If it does not, check back through this section.

Tidying the spreadsheet

Whilst we have the correct answers, the spreadsheet looks rather a mess at present. Some cells cannot be read completely, some of the numbers are money, some are not, and yet they all look the same. The first thing to do is to widen some of the columns to allow the text to be read, and perhaps narrow one too.

Go to the B at the top of that column and click once. This highlights the whole column. If you now move the pointer to the line between columns B and C you will see that it changes shape, and by holding down the left mouse button you can drag to the right and widen the column. Do this for columns B, C, D, G, H, I and J, making them just wide enough to read the text within the spreadsheet. You can use the same idea to make column E much narrower. Column F has now become redundant, so click on the F to highlight it and then go to the **Edit** menu and click on the word **Delete**.

To make the money element more obvious, select cell B4 and click on the currency icon ($). In many cases this will give you dollars rather than pounds, or whatever you want. To change, go to the **Format** menu and click on **Cells...** and then on **Currency** in the new window which appears. Select whichever you want. Do the same for cell G4. Highlight the cells from B9 to D12 inclusive and do the same. Now do it again for cells B15 to D22 inclusive.

Your spreadsheet should now look like Figure 14.5.

Figure 14.5

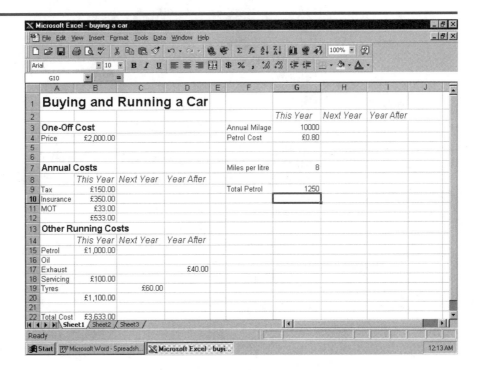

You can continue to change and tidy your spreadsheet to your own requirements, adding colour, more type styles, etc. It is just a matter of how much time you want to spend.

Extending the spreadsheet

Once the basic relationships have been built into your spreadsheet, it is relatively simple to extend these to other, similar, cells. In this case, we want to move the calculations to apply to the following years. We could just type in the formulae again, but it will be quicker and easier to copy the existing ones.

First of all, put in some extra numbers so that your spreadsheet looks like Figure 14.6. These were just the basic data. Now we can copy our formulae.

Go to cell G9 and copy it across to cells H9 and I9. (You can do this either from the **Copy** command on the **Edit** menu, or by taking the pointer to the bottom right corner of the selected cell, so that it changes shape to a small plus sign, holding down the left mouse button, and dragging across to the other two cells.) You will now see the answers for total petrol appear. The original formula was

Figure 14.6

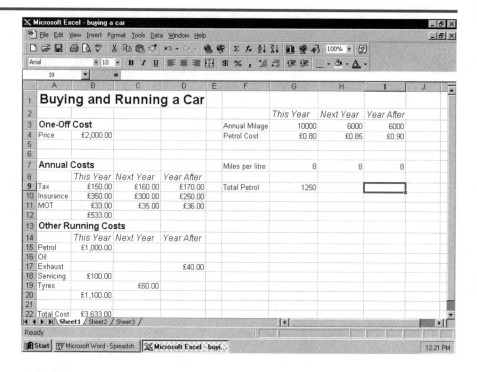

=G3/G7

and if you look at the new cells you will find that in H9 we have

=H3/H7

and in I9 we have

=I3/I7

In other words, when you copy a formula from one cell to another, the **relative references** are maintained.

Now go to cell B15 and copy the formula over to cells C15 and D15 to incorporate the cost of the petrol for each of the next two years.

In cells B12 and B20 we used autosum (Σ) to add the columns, these too can be copied over to the C and D columns to get the sub-totals. Similarly, the addition in cell B22 can be copied to cells C22 and D22.

Finally you can use autosum to add the three year totals together to get the overall cost of the car whilst you are at college.

Your spreadsheet should now look like Figure 14.7.

Using the spreadsheet

The total over the three years is well above the amount you were given, so we can look for ways to reduce it. This will consist of changing individual cells to see the overall effects and is sometimes called a **What if ...** investigation or scenario.

As a first attempt to reduce costs, try reducing the annual milage travelled. Each time that cell G3 is changed, for example, so are cells B15, B20, B22 and F22.

Figure 14.7

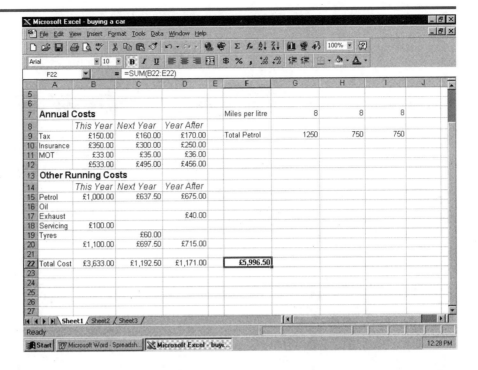

An alternative way of reducing costs would be to buy a more economical car. This would mean that cells G7, H7 and I7 would increase (try 10) and overall cost would go down.

You could always buy a cheaper car in the first place!

What you have just done is a key element of using spreadsheets. Copying formulae makes the constuction fairly quick, but 'finding out what happens if ...' is probably the most effective use of spreadsheets.

Built-in functions   Given time, some experience and imagination, you can make spreadsheets perform almost any kind of calculation. However, why re-invent the wheel? Most of the basic statistical, mathematical and business functions have already been worked out, and many are included with your spreadsheet. Other, more specialist functions can be purchased as add-ins. In fact you have already used the built-in function for AutoSum.

Full use of such functions is beyond this book, but to illustrate the idea, try putting the following numbers into column A of a blank spreadsheet:

10
12
15
12
16
20
16
18
13
11

Go to the **Tools** menu and click on **Data Analysis**. (You may have to add this to the menu on some machines.) Select **Descriptive Statistics** and tell it to find the **Summary Statistics** for the **Input Range** which includes your data. Your results should look like the spreadsheet in Figure 14.8.

Figure 14.8

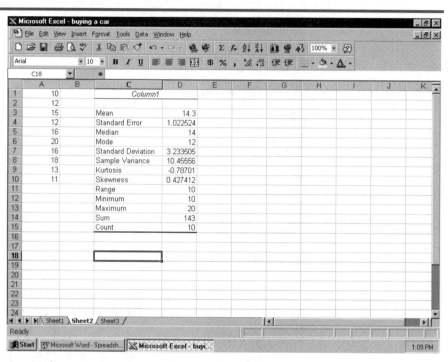

Other, more complex, functions are also included with most spreadsheets.

Obviously any built-in function may not meet your specific needs and so it may be necessary to modify the results, or select only certain parts. Unique situations will always require you to build spreadsheets from scratch!

Section 6

Re-tests

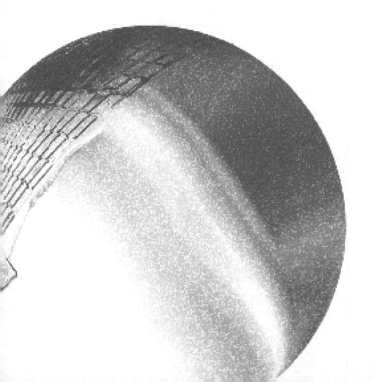

CHAPTER 15

Re-tests

This chapter is divided into four sections, matching the first four sections of the book. After each set of re-tests you will find the answers. Use this section to confirm your learning and understanding after you have worked through at some of the text.

1 Numbers: The re-tests

Now that you have looked through Section One, and done at least some of the exercises, you should try our re-test. Hopefully you will now get a higher score than you did on your first attempts. You can always come back to this section at a later time.

Re-test 1: The basics

1 $12 + 7$

2 $10 - 6$

3 $12 - 6 \times 3$

4 $12 \times 3 - 6$

5 $-9 + 18 \times 2 - 7$

6 $12 \times (-2) + 4 - 8$

7 $11 + (-5) \times (-7) + 3$

8 $\frac{8}{3} - 4$

9 $(12 \div 3) \times 4 + 2 \times 8$

10 $(13 - 5)\,(8 \times 10)\,(-2+4)$

Re-test 2: Working with fractions

1 $\frac{1}{8} + \frac{1}{2} + \frac{1}{4}$

2 $\frac{1}{3} + \frac{1}{3} - \frac{1}{6}$

3 $\frac{1}{16} - \frac{1}{32}$

4 $1\frac{1}{4} \times \frac{1}{2} \times 2\frac{3}{4}$

5 $\frac{1}{8} \times \frac{4}{7}$

6 $\frac{2}{5}$ divided by $\frac{1}{5}$

7 $\frac{2}{3}$ divided by $\frac{4}{5}$

8 $2\frac{1}{4}$ divided by $1\frac{1}{3}$

Re-test 3: Decimals and rounding

1 $1.3 + 0.63$

2 0.4×0.6

3 $0.3 \times 1.4 - 0.4$

4 $1.2 + 1.5 \times 3$

5 $0.25 \times \frac{2}{3}$

Round the number 0.441257624

7 to 5 decimal places

6 to 7 decimal places

8 to 2 decimal places

9 to 1 decimal place

Re-test 4: Percentages and ratios

1 20% of 40

2 10% discount on £19.99 leaves

3 The cost of a particular service was £8 per hour. It has now been subject to an increase of 5% and a further 3%. What is the new cost?

4 Divide £1,080 in the ratio 7:5

5 Three friends win a lottery prize of £600. Payout is in the ratio 3:2:1 between the friends. How much does each get?

Re-test 5: Powers

1 $3 \times 3 \times 3$

2 4^3

3 $4^2 + 4^3$

4 $4^2 \times 4^3$

5 $2^2 + 3^2$

6 $4^5 \div 4^3$

7 $3^2 \div 3^4$

8 $\sqrt{625}$

9 $\sqrt{(4^6)}$

10 $\sqrt[3]{125}$

Numbers: the re-test answers

Re-test 1: Answers

1 19

2 4

3 −6

4 30

5 20

6 −28

7 49

8 $-1\frac{1}{3}$

9 32

10 1280

Re-test 2: Answers

1 $\frac{7}{8}$

2 $\frac{1}{2}$

3 $\frac{1}{32}$

4 $\frac{55}{32}$ or $1\frac{23}{32}$

5 $\frac{1}{14}$

6 2

7 $\frac{5}{6}$

8 $\frac{27}{16}$ or $1\frac{11}{16}$

Re-test 3: Answers

1 1.93

2 0.24

3 0.02

4 5.7

5 $\frac{1}{6}$

6 0.4412576

7 0.44126

8 0.44

9 0.4

Re-test 4: Answers

1 8

2 £17.99

3 £8.65 (rounded)

4 £630 and £450

5 £300:£200:£100

Re-test 5: Answers

1 27

2 64

3 80

4 1,024

5 13

6 16

7 $\frac{1}{9}$

8 25

9 64

10 5

2

Using calculators: The re-test

Now that you have looked through Section Two, and done at least some of the exercises, you should try this re-test. Hopefully you will now get a higher score than you did on your first attempt. You can always come back to this test again at a later time.

Re-test 6: Using calculators

Evaluate each of the following using your calculator:

1 42.3 + 7.9 + 2.1

2 3.57 − 4.2 + 4.852 + 7.631 − 2.4785 − 3.8421

3 $32.483 - \dfrac{7.85}{3.1} + 6$

4 $\dfrac{(32.483 - 7.85)}{(3.1 + 6)}$

5 12.47 × 8.39 × 2.47

6 13% of 62

7 $\dfrac{(12 \times 7521 - 32 \times 451)}{\left(12 \times 475 - 32^2\right)}$

8 $\dfrac{77}{12} - 1.246 \times \dfrac{48}{12}$

9 $\sqrt{17 \times 3 + 45 \times 2}$

10 $\dfrac{356.48}{\sqrt{4592.1 \times 3876.483}}$

Using Calculators: The re-test answers

Re-test 6: Answers

1 52.3

2 5.5324

3 35.950742

4 2.706923076923

5 258.419551

6 8.06

7 16.21471343028

8 1.432666666667

9 11.87434208704

10 0.08449102111921

Algebra: The re-tests

Now that you have looked through Section Three, and done at least some of the exercises, you should try these re-tests. Hopefully you will now get a higher score than you did on your first attempt. You can always come back to this test again at a later time.

Re-test 7: The basics

1 A company has identified two components of costs associated with a particular product: a fixed cost (factors such as rent and heating) of £5,000 per year and a cost associated with making each unit (factors such as materials and labour) of £5 per unit. If c is used to represent total cost and x is used to represent the number of units, then

(a) express the relationship between total cost and the two components of cost by means of an equation,

(b) use this equation (from part (a)) to calculate the total cost for a production level of 10,000 units per year.

2 The cost of club membership is £50 a year with the additional charge of £2.50 for each visit. If c is the total cost (membership plus entry) and x is the number of visits, then

(a) express the relationship between total cost and the number of visits,

(b) use this equation (from part (a)) to calculate the total cost for a person who made 8 visits

By substitution, answer questions 3 to 5.

3 If $x = 17.3$ and $y = 16.2$, then $2x + 3y =$

4 If $a = 4$ and $b = 3$, then $2a + 3ab - b =$

5 If $a = 5$ and $b = -3$, then $2a + 4b^2 =$

6 Calculate -8 add 16 subtract -4 add -10

Simplify the expressions given in questions 7 to 10.

7 $2 \times x \times x \times x + x \times y \times z - x \times x + 2x^3 - x^2$

8 $2ab + a \times a \times a - a \times b + z \times z$

9 $\dfrac{x^3 y^2 z}{xz}$

10 $a^4 b^{-2} a^{-2} b^3$

Re-test 8: Powers, brackets and order

Simplify the following expressions:

1 $6y^2 + 4y^2 - y - 3 \times y \times y$

2 $2a^2 - 3b^2 + 2a^2 + b^2 - 4a^2$

3 $2x^{-2} \times 4x^{-4}vy^{-1} \times x \times y$

4 $3b^4 \times 2b^{-2} \div b^{-1}$

5 $\left(3ab^{-2}\right)^3$

6 $\sqrt[3]{a^3 b^6 c^9}$

7 $\left(\sqrt[3]{x^6 y^6}\right)^{\frac{1}{2}}$

8 $(a + 2b + 3c) \times 3 - a - b - c$

9 $2x \times 4 - 3(2x - y)$

10 $4y \times 3 \times 3y - 3(2 + 4)(x + y)$

Algebra: The re-test answers

Re-test 7: Answers

1 (a) $c = 5000 + 5x$
where 5000 is the fixed cost (which does not vary with x), and
$+5x$ is the variable cost (which increases with x)

(b) Let $x = 10000$
$c = 5000 + 5 \times 10000$
$c = 5000 + 50000$
$c = 55000$

2 (a) $c = 50 + 2.50x$
where 50 is the fixed cost, and $2.50x$ is the variable cost.

(b) Let $x = 8$
$c = 50 + 2.50 \times 8$

$c = 50 + 20$
$c = £70$

3 $34.6 + 48.6 = 83.2$

4 $8 + 36 - 3 = 41$

5 $10 + 36 = 46$

6 2

7 $-2x^2 + 4x^3 + xyz$

8 $a^3 + ab + z^2$

9 $x^2 y^2$

10 $a^2 b$

Re-test 8: Answers

1 $7y^2 - y$

2 $-2b^2$

3 $8x^{-5}$ (remember $y^0 = 1$)

4 $6b^3$

5 $27a^3 b^{-6}$

6 $ab^2 c^3$

7 xy

8 $2a + 5b + 8c$

9 $2x + 3y$

10 $36y^2 - 18x - 18y$

4

Equations and graphs: The re-tests

Now that you have looked through Section Four, and done at least some of the exercises, you should try these re-tests. Hopefully you will now get a higher score than you did on your first attempts. You can always come back to these tests again at a later time.

Re-test 9: The basics

1 It has been decided to review coursework marks in a more systematic way by taking a sample and double checking the marking. The sample size (s) is to be determined by taking the square root of the number of assignments submitted (n) and rounding as appropriate. Show how the sample size is to be determined by means of an equation.

2 A local authority uses 2 types of bus, referred to as A and B. A can carry 30 people and B can carry 45 people. Use an equation to show how many people (P) can be carried by a combination of buses. If the cost of a seat on A is 65 pence and the cost of a seat on B is 58 pence, use an equation to show the cost of using a combination of buses (C).

3 The volume of liquid produced by a particular process in mlr (v) is given by:

$$v = 16 + 3.2t + 1.2t^2 + 0.8t^3$$

where t is time in seconds.

Find the volume of liquid produced after 8 seconds.

Solve the following equations:

4 $-8x = -2x - 42 \times x + x - 105$

5 $4b(b - 2) + 3(b - 4) = 4b^2 - 6(b - 5)$

Re-test 10: Graphs

Draw appropriate axes that will allow you to plot points A to E as defined below, and follow the instructions given in questions 1 to 3 and answer questions 4 and 5.

1 Draw a line from A (−5,−3) to B (2, 4)

2 Draw a line from B (2, 4) to C (4,−2)

3 Draw a line from C (4,–2) to A (–5,–3)

4 Does the point D (3,–2) lie within the shape produced by questions 1 to 3?

5 Does the point E (–2, 1) lie within the shape produced by questions 1 to 3?

6 Draw a graph to show how the number of enquiries received by an organization has changed over time given the following information:

Year	Number of Enquiries
1993	83
1994	78
1995	64
1996	62
1997	67
1998	74
1999	79

Comment on the shape.

7 The sales of a particular travel guide have been recorded as follows

Year	Quarter 1	Quarter 2	Quarter 3	Quarter 4
1997	80	122	130	77
1998	82	118	129	82
1999	83	120	130	

Draw an appropriate graph to show how sales have changed over time.

8 Given the information below, plot the number of accidents against the number of employees (in thousands):

Year	Number of employees (in thousands)	Number of accidents
1	1	50
2	2	50
3	4	30
4	7	20
5	9	10

Comment on the graph you have drawn.

Re-test 11: Linear equations

Draw appropriate axes and on the same graph plot the following:

1 $y = 0.5x$ between $x = -20$ and $x = 40$

2 $x = 10$ between $y = -25$ and $y = 5$

3 $y = -20 - 0.5x$ between $x = -20$ and $x = 40$

4 Using appropriate axes, plot the following points:

x	y
5	–225
6	–280
8	–390

5 On the graph you have just drawn (question 4), join the points with a straight line. Now determine the linear equation that describes the line.

Re-test 12: Simultaneous equations

1 Given the following equations

$$y = 4x - 55$$
and $$y = -2x + 35$$

(a) plot both on the same graph and identify the point of intersection

(b) solve as simultaneous equations

Solve each of the following pairs of simultaneous equations using appropriate algebra:

2 $2y + 3x = 34$
$7y + 5x = 53$

4 $-3y - 3x = -36$
$y + 4x = 33$

3 $-4y + x = 13$
$2y - x = -5$

5 $4x - 2y = 1$
$-3x + 7y = 13$

6 A company makes two types of frame, x and y. Each frame x requires 1 hour of labour and 6 litres of moulding material, whereas each frame y requires 2 hours of labour and 5 litres of moulding material. The total number of labour hours available each week is 40 and the total amount of moulding material available each week is 150 litres. Determine how many frames that can be made that will fully utilize labour time and moulding material.

Re-test 13: Quadratic equations

1 Draw the graph of $y = x^2 + 2x - 35$ from $x = -10$ to $x = 10$ and identify the roots.

Using the method of factorization, find the roots of the following quadratic equations:

2 $x^2 + 3x - 70 = 0$

4 $x^2 - 1 = 0$

3 $x^2 + 4x = 0$

5 $x^3 + 5x^2 - 24x = 0$

Using the equation method solve the following quadratic equations

6 $x^2 + 3.42x + 2.7392 = 0$

7 $x^2 - 1.3225 = 0$

Equations and graphs: The re-test answers

Re-test 9: Answers

1 $s = \sqrt{n}$

2 $P = 30A + 45B$
$C = 0.65 \times 30 \times A + 0.58 \times 45 \times B$
$C = 19.5A + 26.1B$

3 $v = 16 + 3.2 \times 8 + 1.2 \times 64 + 0.8 \times 512$
$= 16 + 25.6 + 76.8 + 409.6$
$= 528 \, mlr$

4 $-8x = -43x - 105$
$35x = -105$
$x = -3$

5 $4b^2 - 8b + 3b - 12 = 4b^2 - 6b + 30$
$b = 42$

Re-test 10: Answers

The graph required for questions 1 to 5 is shown below:

4 Point D lies within the shape as shown

5 Point E lies outside the shape as shown

Figure 10.1

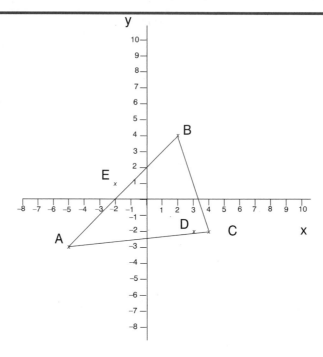

6

Figure 10.2

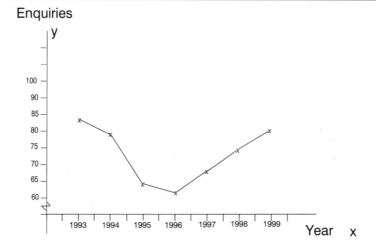

The annual number of enquiries drops from 1993 to 1996 and then begins to increase again. A curve would describe this relationship more effectively than a straight line. A researcher would need to look at other factors to try to explain this relationship.

7

Figure 10.3

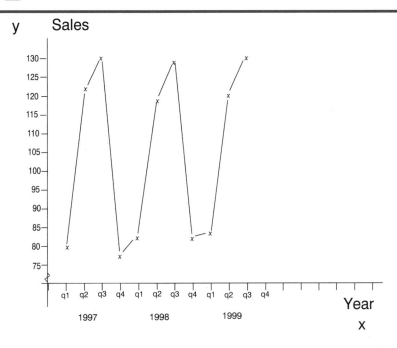

No general trend but clear seasonal variation with quarter 3 and then quarter 2 predictably higher.

8

Figure 10.4

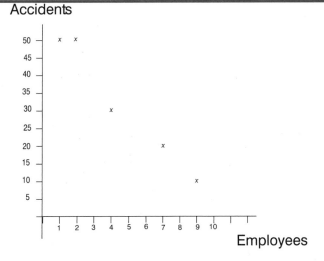

The graph suggests that as the number of employees increases the number of accidents decreases (we would always like to see a decrease in the number of accidents but this is unlikely to be achieved just by an increase in the number of employees!). If we look more closely at the data, the number of employees has increased over time and the number of accidents has decreased over time. It could be that we are looking at the effects of a third or indeed fourth variable. It could be that as the organization has grown, it has improved its safety procedures or that over time training has improved. All we need to note is that a simple graph may not provide the answers and that without careful interpretation, the presentation of findings can be misleading.

Re-test 11: Answers

The graph required for questions 1 to 3 is shown below:

Figure 11.1

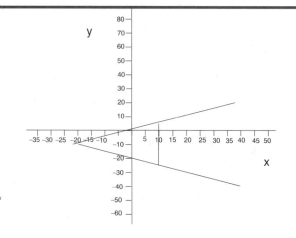

Did you get the A on its side? Again just check the corner points for correctness.

The graph below (Figure RT11.2) shows the plotting of points from question 4 and the determination of the equation $y = 50 - 55x$.

Figure 11.2

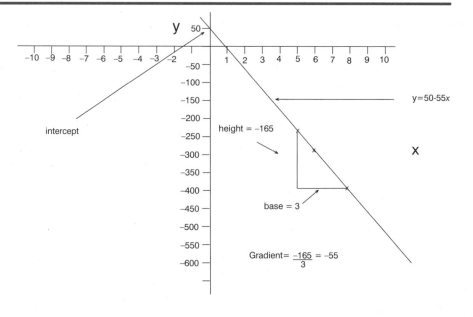

Re-test 12: Answers

1 (a)

Figure RT12.1

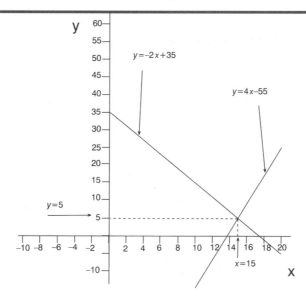

(b) $x = 15$ and $y = 5$

2 $x = 12$ and $y = -1$

3 $x = -3$ and $y = -4$

4 $x = 7$ and $y = 5$

5 $x = 1.5$ and $y = 2.5$

6 We can express this problem in the form of two equations, one for the labour requirements and one for materials:

Labour: $x + 2y = 40$

Materials: $6x + 5y = 150$

This can be represented graphically:

Figure RT12.2

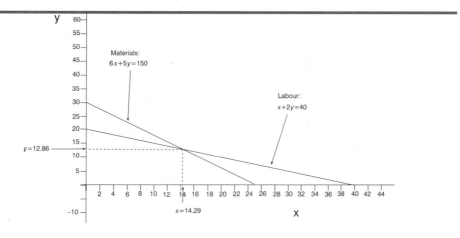

The answer is $y = 12\frac{6}{7}$ and $x = 14\frac{2}{7}$

(This problem is more fully considered in *Quantative Methods for Business Decisions* by Curwin and Slater.)

Re-test 13: Answers

1 The roots are $x = -7$ and $x = 5$

Figure RT13.1

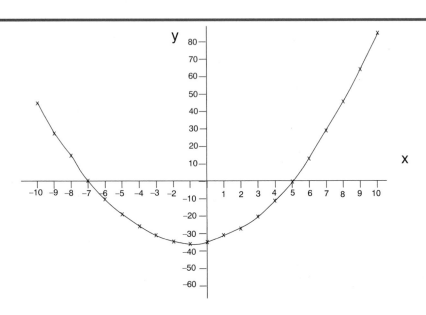

2 The roots are $x = -10$ and $x = 7$

3 The roots are $x = -4$ and $x = 0$

4 The roots are $x = -1$ and $x = 1$

5 The roots are $x = -8$, $x = 0$ and $x = 3$

You should note that this is **not** a quadratic equation as defined by the general form (in terms of x^2 and x) but is included as a more challenging exercise in factorization.

6 $a = 1$ $b = 3.42$ $c = 2.7392$

roots of the quadratic are $x = -2.14$ or $x = -1.28$

7 $a = 1$ $b = 0$ $c = -1.3225$

roots of the quadratic are $x = -1.15$ or $x = 1.15$

Guildford College
Learning Resource Centre

Please return on or before the last date shown
This item may be renewed by telephone unless overdue

Class: _____ 510 CUR _____

Title: _Improve Your Maths_____

• Author: _CURWIN, Jon & SLATER, Roger_